Identification of North American Commercial Pulpwoods and Pulp Fibres

Identification of North American Commercial Pulpwoods and Pulp Fibres

I. STRELIS
B.Sc., F.R.M.S.

and

R. W. KENNEDY
Ph.D.

Published in association with the
Pulp and Paper Research Institute of Canada by
University of Toronto Press

© University of Toronto Press 1967
Printed in Canada
Reprinted in 2018
ISBN 978-1-4875-7358-4 (paper)

PREFACE

BECAUSE OF its availability in large quantities, wood is the main fibrous raw material of the paper industry. The various woods are usually classified into individual types or groups according to their paper-making characteristics. In determining the quality of a paper product, the inherent properties of wood are as important, if not more so, than the pulping processes and other mill treatments to which the wood is subjected. Consequently, it is commercially rewarding to be able to select the right type of wood for the particular end product required. The ability to identify wood is therefore of considerable practical importance.

In order to determine what material has been used in a given product, it is also necessary to have means of identifying a wide variety of different fibres, of both woody and nonwoody origin. Although most paper is made from wood, a large number of fibres, of origin other than wood, find their way into paper products. Such fibres are used because of their local availability or because certain specialty papers call for their use.

Mill analysts often need guiding references on how to proceed with the identification of both wood and fibres. In many cases this information is available but scattered through many publications, and frequently it is presented in a form that is too general to be of use. This publication has tried to overcome these difficulties by bringing together what are considered the most useful wood and fibre features required for identification in the paper industry. It is felt that the many photomicrographs illustrating these features will prove to be an invaluable visual aid. An attempt has been made to present the information in as systematized a form as possible. It is realized, however, that not everything will be readily identified because of the great similarities that exist between many wood structures and individual fibre features. To make full use of this publication, some experience as an analyst is required. At the same time it may well prove to be useful as a self-training manual.

Finally, it should be mentioned that the preparation of this publication has been undertaken at the suggestion of the Testing Methods Committee of the Technical Section, Canadian Pulp and Paper Association. The Committee felt that there was a great need for a publication of this type.

I. S.

Pointe Claire, Quebec, 1967

CONTENTS

PREFACE v

PART ONE: IDENTIFICATION OF WOOD AND WOOD FIBRES

Introduction	3
Identification of Wood	5
Key for the Identification of Commercial Pulpwoods and Their Pulps	12
Identification of Wood Fibres	28

PART TWO: IDENTIFICATION OF NONWOODY AND MAN-MADE FIBRES

Introduction	43
Fibre Identification	44
Seed Hairs: Fibres	48
Bast Fibres	55
Leaf Fibres	65
The Pulps of Grasses	72
Animal Fibres	82
Mineral Fibres	87
Organic Man-Made Fibres	90

CONTENTS

APPENDIXES

I.	Solubility of Fibres	106
II.	A Scheme of Fibre Analysis by Solvent Action	108
III.	Colour Reactions of Fibres to Selective Staining	109
IV.	Identification Key to Nonwoody and Man-Made Fibres	111

INDEX 115

PART ONE

IDENTIFICATION OF WOOD
AND WOOD FIBRES

I. Strelis and R. W. Kennedy

INTRODUCTION

THE RAW MATERIALS used in the production of paper come from many sources, but of these wood is the most important. There is a great variety of wood types that correspond with the many tree species that are to be found. These woods differ in their anatomical structure and in their physical and chemical properties. Wood fibre properties together with pulping processes and other mill treatments determine the quality of paper and paper products. Consequently, the ability to recognize the different kinds of wood is of importance (1) in meeting the pulp requirements for the manufacture of various grades of paper, (2) for determining the components of a particular type of a paper, and (3) for general information.

WOOD IDENTIFICATION

Trees may be identified by their botanical features, but these are no longer available with wood samples. Those who have had long experience in wood identification are often able to distinguish types of wood by their general appearance and by some of their physical properties such as colour, lustre, odour, taste, weight, and hardness. The reliability, however, of this means of identification is very much in doubt, because colour, as well as some of the other properties, will change with seasoning or decay. In addition, qualitative description of these features is difficult and cannot be communicated adequately either in words or by photographs.

A more reliable means of identification is offered by the gross structural differences and the variability in cell patterns and cell features which occur in the various woods. These variations in characteristics provide the microscopist with diagnostic features by which the wood of one genus may be distinguished from that of another.

It is the purpose of this part to present a systematized method, in the form of a key, for the identification by genus of all Canadian commercial pulpwoods and their fibres. This key has been prepared by using features which can be readily recognized, photographed, and defined. Using the key for identification often requires the observation of both the macro features of the wood block, using a hand lens, and the micro features to be found in thin sections, examined under the microscope. The adaptation of the key to an edge-notched card system is also described.

FIBRE IDENTIFICATION

Mill analysts are mainly concerned with identifying the constituents ("fibres") of pulp or paper. The difficulties which arise in the identification of fibres are greater than those occurring with wood. In pulps, the anatomical relationship of the cells to each other no longer exists and thus cannot be used as a diagnostic feature. Consequently, identification of pulp elements depends solely upon the very few features to be found in individual cells. Mechanical treatment of pulps obscures even further the features required for recognition.

Although the key mentioned earlier will be of some use in fibre recognition, the commonest means of identifying pulp fibres is by comparison of unknown fibres with either fibres of known samples or with authentic photomicrographs. Photomicrographs and descriptive information of typical fibre features as an additional aid to identification are included in succeeding pages.

IDENTIFICATION OF WOOD

SOFTWOODS AND HARDWOODS

For practical purposes woods are grouped into two main classes, namely, softwoods and hardwoods. These two terms, however, are not very definitive since some softwoods are harder than certain hardwoods.

In North America the term "softwoods" means the conifers, which are cone-bearing trees, having needle-like, or scale-like leaves. They are all evergreens with the exception of larches which shed their leaves in the autumn.* Softwood trees belong to the order coniferales of the Gymnosperms. The vertical structure of softwoods is composed largely of cells similar in shape. These cells, known as tracheids, are tube-like with closed and pointed ends and are arranged parallel to the grain. A cut across any of the softwoods reveals that the wood cells are in radial rows which extend from pith to bark in essentially straight lines. As these cells are of small and relatively uniform diameter and are invisible to the naked eye, softwoods are also known as non-porous woods.

The term "hardwoods" refers to the broad-leaved trees, which are all deciduous. Botanically, they belong to the dicotyledon class of the Angiosperms. The principal vertical structure of hardwoods is composed of relatively long, fibrous cells of small diameter, as well as shorter, larger diametered cells called vessels. In a cross-section of the wood the vessels appear as numerous rounded pores which in some species are visible to the naked eye. Consequently, the hardwoods are also known as the porous woods.

There generally is no difficulty in distinguishing softwoods from hardwoods. Preliminary examination of the cross-section of any wood with a hand lens will place it in one of the two groups. (Compare, photomicrographs SF 1 and SF 21 with HF 1 and HF 2, pages 13, 17, and 19.)

Softwoods and hardwoods are slightly different in chemical composition and react differently with certain chemicals. This difference provides a method for

---

*Bald cypress, *Taxodium distichum*, growing in swamps in the southern United States, is also a deciduous conifer, but it is not used for pulp.

distinguishing the two types of wood by chemical means. For this purpose, a fresh one per cent solution of potassium permanganate is prepared. The wood is immersed in it for two minutes and then directly transferred into 12 per cent hydrochloric acid for one minute. From here it is transferred into 1 per cent or even concentrated ammonium hydroxide. The softwoods always turn yellow, while the hardwoods take on a colour varying from pink to purple.[1]

ABILITY TO IDENTIFY WOOD

The question is often asked whether or not it is possible to identify a given wood species with certainty. The answer is that the identification of different wood genera is usually possible, but that distinguishing individual species is sometimes difficult and in many cases impossible, because species of the same genus are so closely related in anatomical structure. In view of this, the key that follows represents an approach to identification of wood by genera or by dissimilar groups of species within the same genus. In this way the key avoids the debatable details of specific identification. The degree of uncertainty associated with identification of individual species can be estimated only after considerable experience in problems of identification and cannot be properly communicated in a written key. Further information on the problems of identification may be obtained from standard references.[2,3]

Experience

It should be emphasized that ability to recognize morphological features, a prerequisite for identification, depends upon a person's familiarity with wood and its anatomical structure. It is important that he know how to prepare and examine wood and what to look for. Anyone without previous training in wood anatomy should train himself on authentic wood specimens. In this connection it is useful to keep a sample collection of all Canadian woods for reference purposes. Such samples are sold by the Forest Products Laboratories, Ottawa.

Origin of Sample

It is always an advantage to know the locality from which an unknown sample has been obtained. For example, there is always the possibility of confusing eastern hemlock with western hemlock. If it is known, however, that a sample comes from the forests of eastern Canada, it can be said with assurance that it is eastern hemlock.

By taking into account the commercial range of distribution by species (see diagnostic features), it is possible to reduce the number of species to be considered when identifying wood samples.

Examination of Wood

When assessing wood by general appearance, it may be found that even woods of different genera appear very similar to the naked eye. This is particularly true with softwoods, which are the main source of pulpwood. It also is occasionally difficult to distinguish softwoods from hardwoods by general appearance.

Woods are distinguishable by their macro- and micro-morphological features. These features may be observed in three different planes: (1) the cross-section or transverse plane, produced by cutting across the wood perpendicular to its grain; (2) the radial plane, exposed by cutting parallel to the grain along a ray; and (3) the tangential surface, revealed by cutting parallel to the grain but at a tangent to the growth rings and thus perpendicular to the rays. These planes are abbreviated in the list of diagnostic features as x, r, and t, respectively. As most features are not visible to the naked eye, they need to be observed more closely under magnification.

In wood identification, two main methods of examination are employed. The first covers the macroscopic examination of the various gross or macro features and requires the use of a hand lens of at least 10x magnification. The second method is employed to observe the finer or micro features which usually are not visible under a hand lens. This method requires the preparation of thin wood sections which can be examined under a microscope, at a magnification of between 25x to 400x, using transmitted light. Macro examination will seldom lead to positive identification, and the combined observations by both methods are generally required.

The Key

Useful diagnostic features for softwoods and hardwoods have been numbered, defined, and, in most cases, illustrated. Separate lists are given for both softwood and hardwood genera and species considered to be of importance to the pulp and paper industry. The numbers accompanying each genus, species, or group of species refer to the diagnostic features categorizing that particular genus or group. A sample is identified by determining the numbered features present and finding in the lists the genus (or group) having all those features. A description of transcribing the key to an edge-notched card filing system is also included.

MACROSCOPIC EXAMINATION OF WOOD

The cross-surface of the wood is the most revealing for this kind of examination, and consequently it needs to be properly exposed. A strong, sharp knife should be used to make a straight, smooth cut. If a dull knife is used, it tears and crushes the cell walls and distorts the cell shapes. This interferes with the observations of the cell size, shape, and the arrangement of the cells in the tissue, and may lead to an inaccurate diagnosis.

In many cases it is very helpful to moisten the freshly cut cross-surface with a drop of water immediately before examination. Often it helps to bring out certain minute features, but in a few other cases it may hinder observations. Both ways should be tried. In addition, visibility of features on a wood surface often depends upon illumination and the angle at which lights falls upon it.

Softwoods

Normal resin canals, both longitudinal and transverse, are present in the woods of four softwood genera: pines, spruces, larches, and Douglas fir. They are not present in the other softwoods: true firs, hemlocks, and cedars. Examination for these canals thus is an important first step in analysis of unknown coniferous woods.

Normal resin canals vary in diameter (see softwood features 1 to 8), depending upon species. They are always visible under the hand lens in cross-surface of softwood, appearing as scattered dots or white and black flecks. They are more pronounced in pines and less so in the other three genera.

When a softwood tree is injured, it develops *traumatic* resin canals in the wound tissue. Because of this, the wood of the true firs and hemlocks, although not that of the cedars, will have traumatic resin canals in such areas. Therefore, one should avoid being misled by the presence of traumatic canals which are arranged in a tangential row, unlike the random distribution of normal canals. These tangential rows (parallel to annual rings) can be up to one inch in length or even longer.

In general, the hand lens helps to differentiate softwoods into those with and without resin canals and also species with large canals from those with smaller ones. In addition, a number of features can be observed with the naked eye: earlywood—latewood transition, occurrence of dimpled grain on tangential surface, colour, lustre, and the presence or absence of heartwood. These features, along with odour, allow the experienced observer to make a tentative conlusion as to genus. However, since softwoods have a much more uniform anatomical structure than hardwoods, positive identification will often depend on microscopic examination.

Hardwoods

In all hardwoods the pores are big enough to be observed with a hand lens on cross-section. Because the diameters of these pores, as well as their grouping and distribution in the annual rings, are very characteristic in each genus, they are of great diagnostic value (see hardwood features 1 to 3).

Another noticeable feature in a cross-section is parenchyma, although not all hardwoods have it. When parenchyma is abundant, it often gives rise to very definite patterns on the cross-section (see hardwood features 11 to 13). They stand out as lighter coloured tissue against a darker background. Certain patterns are typical for particular hardwoods and are of real diagnostic value. They are best observed with a hand lens.

In many hardwoods, the third feature that is readily detectable is the system of rays (see hardwood features 16 to 23). In woods of oak, beech, and maple the larger rays are visible to the naked eye on all three planes of view. They appear as a series of radiating lines on the transverse surface, while on freshly split radial surfaces they are visible as a series of streaks. The end-view of the ray appears on the tangential surface as a short (maple) or long (oak) line. Hardwood rays are much more variable in width and in height than those of softwoods.

Further, with some experience, most hardwoods can be identified with a hand lens on the basis of their pronounced rays, distribution of pores and parenchyma, and colour.

MICROSCOPIC EXAMINATION OF WOOD

In many wood species most of the features can be observed only while examining wood sections under the higher magnification of a microscope (25x to 400x). For this purpose the wood has to be adequately prepared and sectioned in all three planes: transverse, radial, and tangential. As microscopic examination is made with transmitted light, these sections need to be thin enough to be translucent.

Wood samples for identification are submitted in many forms: logs, discs, blocks, boards, chips, matchsticks, etc. Freshly felled woods, especially softwoods, are best for sectioning. However, submitted samples are often seasoned and air-dry. Wood in this condition is too hard for sectioning.

Dry wood needs to be softened before it can be sectioned. There are several ways to do this. The simplest method is boiling in water until the block sinks when placed in cold water. If the sample is large, the first step is to reduce it in size, care being taken to include at least one annual ring of full width. The smaller the block, the less is the time required for softening and water-logging. For sectioning, blocks of approximately $1 \times 1 \times 1.5$ cm are adequate. This method of boiling is usually adequate in preparing the softwoods and softer hardwoods for sectioning. It is useful to remember that the wood will be softer if it is still hot from the boiling water or is kept hot by a guided jet of steam than if it is allowed to cool before sectioning.

The harder woods will always need more softening before they can be sectioned. Some of the softening solutions are corrosive and require a prolonged treatment time. The simpler wood softening methods are as follows:

(a) Prolong the period of boiling in water. Depending upon wood hardness, the period required may be from a number of hours to several days under a reflux condenser.

(b) Reflux the wood in a solution of potassium hydroxide (2 to 4 per cent) in 95 per cent alcohol for one to two hours. Then wash it thoroughly in running water.

(c) Reflux the wood for one hour in a mixture of glacial acetic acid and 6 per cent hydrogen peroxide made up at the ratio of 1:2. Then wash the wood thoroughly in running water and neutralize the acid by boiling the wood for a few minutes in water with traces (a pinch in 250 ml) of dissolved sodium carbonate.

Wood is sectioned on a special tool known as a microtome. This tool is usually used only when permanent mounts for further studies are required. For routine identification freehand sections are adequate and are prepared by cutting thin slices with a rigid razor blade (such as a single-edge blade) or a very sharp knife. With a little patience and practice one may cut fine freehand longitudinal sections of wood between 30 and 60 microns thick. Transverse freehand sections are more difficult to prepare than longitudinal sections, but even if the sections are not perfect, the necessary diagnostic features can be observed. Moreover, most of the features requiring microscopic observation are found in the longitudinal sections.

For examination under the microscope a section is placed on a micro slide. A few drops of glycerine-alcohol mixture (1:1) are added on the section, and it is then covered with a coverglass. This slide is next heated on a hot plate to expel air bubbles from the section and the mountant. After the slide has cooled the coverglass is pressed down on the section. The excess of the mountant, which comes out around the edges of the coverglass, is absorbed with blotting paper. Water can also be used as a temporary mountant, although it evaporates rapidly leaving the wood section dry and opaque.

The contrast of unstained wood sections is somewhat low, but often adequate. Thin sections illuminated with high intensity light show even less contrast, and consequently light intensity needs to be controlled. Depth of field is even more important, but this can be adjusted by closing or opening the iris diaphragm of the substage condenser on the microscope. By doing so it is possible to arrive at a setting where the various details in the section can be properly focused for optimum observation. These settings will be different for objectives of different magnifications.

The contrast of wood sections can be increased by treating them with a phloroglucinol-hydrochloric acid solution. A suitable solution may be prepared by dissolving 1.0 gm of phloroglucinal into 25 ml of 50 per cent ethyl alcohol acidified with 10 ml of concentrated hydrochloric acid. A few drops of this solution put directly on a section will cause it to turn red within thirty seconds because of the reaction with lignin. The section may then be examined directly, or mounted in glycerine or water. The stain is not permanent and will fade noticeably within a few hours. The phloroglucinol solution should be freshly prepared at approximately monthly intervals for best results.

Contrast of wood sections can also be increased by staining them in solutions of various basic dyes. The best dye solution is that of safranine O, made up of 0.5 to 1.0 per cent dye in 50 per cent alcohol or in distilled water. The sections are immersed in the dye solution for a few minutes, care being taken not to overstain, then they are washed free of excess dye and mounted as described earlier. Heating of the slide should be omitted in order to avoid leaching of the dye into the mountant.

IDENTIFICATION OF WOOD FIBRES IN PULPS

A wood sample is identified by taking into account all the diagnostic features that have been listed in this report. In contrast to this, when identifying the constituents of wood pulps, many of the features useful in wood identification may no longer be of value owing to dissociation of the cells. Resin canals and earlywood–latewood transition in softwoods, parenchyma patterns, and ray width in hardwoods are examples of such features. Consequently, special attention must be focussed on the type of cross-field pits on the softwood fibres (which actually are longitudinal tracheids), and the nature of the ray tracheids, if present, must be studied to effect identification. Even when this has been done, species separation will be incomplete. For example, the southern pines—jack, lodgepole, and ponderosa—all share the identical features of pinoid ray crossing pits and dentate ray tracheids and thus defy separation in pulp form.

In the case of hardwood pulps, the vessel elements must be studied. If the pulp has been derived from ring-porous hardwood, the largest vessel elements, although scarce in distribution, may be at least 160 to 180 microns in diameter, whereas those of the latewood will be only one-third or less of this diameter. Perforation plates, spiral thickenings, and patterns of pitting must be noted on these elements.

However, many analysts who are concerned with the identification of fibres only and have never needed to learn the key procedure for identification of wood may find use of the key too complicated for their requirements. They may find it simpler, faster, and more convincing to make fibre identification by comparing unknown fibres with fibre photomicrographs and authentic fibre samples, especially if they are involved with only a limited number of possible choices of species. To facilitate wood fibre identification by the *comparison method*, information and illustrations depicting fibre features are provided as visual aids in the section "Identification of Wood Fibres."

KEY FOR THE IDENTIFICATION OF COMMERCIAL PULPWOODS AND THEIR PULPS

DIAGNOSTIC FEATURES FOR SOFTWOODS

ABBREVIATIONS: *x*—cross-section; *r*—radial section; *t*—tangential section; *pulp*—pulp fibre.

(See also illustrations of these features which are designated by the same numbers.)

I. Resin Canals

1. LONGITUDINAL NORMAL PRESENT (*x*). Intercellular spaces scattered throughout growth rings. With hand lens, larger ones are visible as dark cavities, while smaller ones appear as scattered white dots.
2. LONGITUDINAL NORMAL ABSENT (*x*). No intercellular spaces visible.*
3. EPITHELIUM THICK-WALLED (*x*,* *t*). Cells immediately surrounding resin canal thick-walled.
4. EPITHELIUM THIN-WALLED (*x*, *t*). Cells immediately surrounding resin canal thin-walled (often badly torn in sectioning).*
5. AVERAGE DIAMETER OF LONGITUDINAL CANAL IS LESS THAN 70 MICRONS (*x*). Diameter measured in direction parallel to growth rings, and includes epithelium.*
6. AVERAGE DIAMETER OF LONGITUDINAL CANAL ABOUT 90 MICRONS (*x*). Diameter measured in direction parallel to growth rings, and includes epithelium.*
7. AVERAGE DIAMETER OF LONGITUDINAL CANAL ABOUT 120 MICRONS (*x*). Diameter measured in direction parallel to growth rings, and includes epithelium.*
8. AVERAGE DIAMETER OF LONGITUDINAL CANAL IS GREATER THAN 120 MICRONS (*x*). Diameter measured in direction parallel to growth rings, and includes epithelium.*

II. Rays

9. RAY TRACHEIDS REGULARLY PRESENT (*r*). Cells often confined to margins of rays; recognized by their small bordered pits.
10. RAY TRACHEIDS ABSENT (*r*).*
11. RAY TRACHEIDS DENTATE (*r*, *pulp*). Tooth-like projections on horizontal walls of ray tracheids.
12. RAY PARENCHYMA END WALLS NODULAR (*r*, *pulp**). Bead-like projections present on (vertical) end walls of ray parenchyma cells.
13. RAY PARENCHYMA END WALLS SMOOTH (*r*, *pulp**). Very thin, smooth (vertical) end walls on ray parenchyma cells.

*Not illustrated.

III. Cross-field Pits

Pits in the field area delimited vertically by an *earlywood* longitudinal tracheid and horizontally by a ray parenchyma cell.

14. FENESTRIFORM (*r*, *pulp*). 1 to 2 rectangular window-like pits per field.

15. PINOID (*r*, *pulp*). 2 to 6 pits per field. Shapes varying from ellipses to irregular polygons. Borders may or may not be visible around apertures. If visible, often wider on one side of aperture than the other.

16. PICEOID (r, *pulp*). 2 to 6 pits per field. Apertures slit-like, often extending beyond visible border.

17. CUPRESSOID (r, *pulp*). 1 to 6 pits per field. Apertures slit-like, but contained within boundaries of border. Width of apertures slightly less than width of border on either side of aperture.

18. TAXODIOID PITS (*r*, *pulp*). 1 to 4 pits per field. Pits have regular elliptical shape, and larger aperture than piceoid or cupressoid (i.e., aperture definitely wider than border on either side of aperture).

IV. Miscellaneous

19. SPIRAL THICKENING IN EARLYWOOD TRACHEIDS (*r*, *pulp**). Spring-like helix running entire length of longitudinal tracheids.

20. LONGITUDINAL PARENCHYMA IN BODY OF RING (*x*). Cells usually recognized by darkened contents in lumina. Cells interspersed throughout growth ring, and *not* confined to the extreme terminal layer.

21. EARLYWOOD–LATEWOOD TRANSITION ABRUPT (*x*). A sharp, well-defined boundary between the earlywood of a growth ring and the latewood of the same ring.

22. DIMPLED GRAIN (*t*). Indentations visible on split surface, especially when specimen tilted to allow light to be reflected at varying angles from the wood surface.

V. Commercial Range (see map, page 22)

23. EASTERN CANADA–NORTHEASTERN UNITED STATES.
24. SOUTHEASTERN UNITED STATES.
25. WESTERN CANADA–WESTERN UNITED STATES.

ILLUSTRATIONS OF SOFTWOOD FEATURES

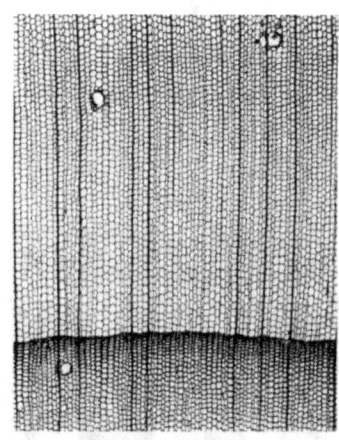

SF 1. Normal resin canals present (*x*, 25x) *Picea glauca*, white spruce.

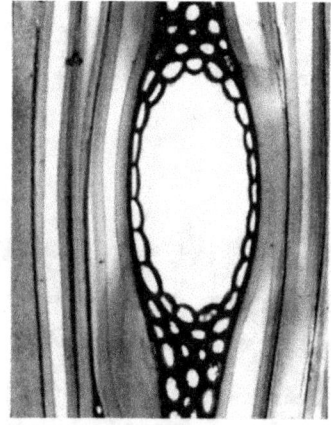

SF 3. Epithelium of resin canal, thick-walled (*t*, 250x) *Larix occidentalis*, western larch.

*Not illustrated.

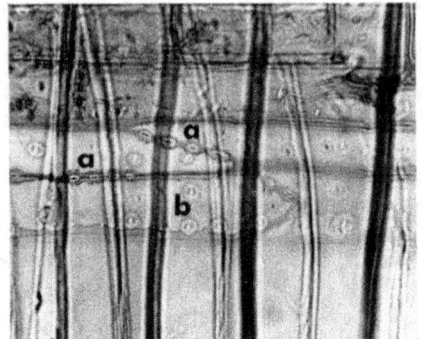

SF 9. Ray tracheids present (r, 400x) *Picea rubens*, red spruce: (a) sectional view of small bordered pits between two ray tracheids; (b) surface view of small bordered pits of a ray tracheid.

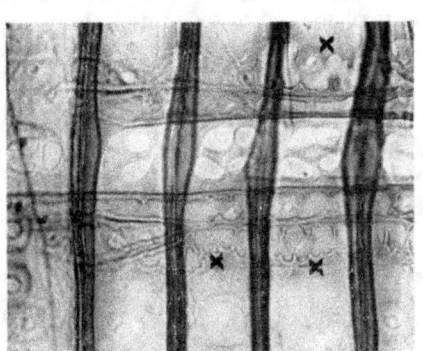

SF 11A. Ray tracheid dentate (x in figure) (r, 400x) *Pinus* spp., hard pine.

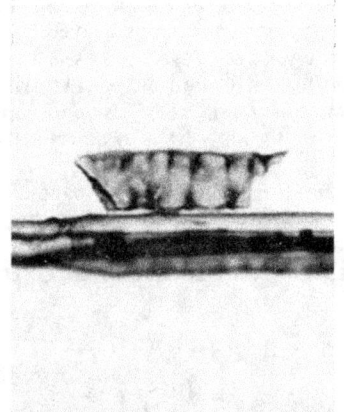

SF 11B. Ray tracheid dentate (pulp, 400x) *Pinus banksiana*, jack pine.

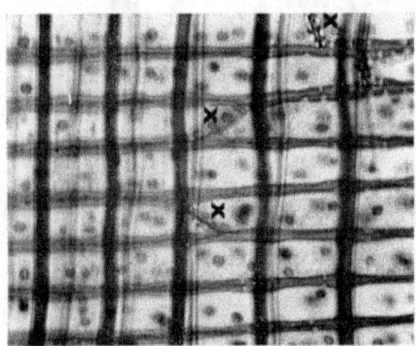

SF 12. Ray parenchyma end walls nodular (x in figure) (r, 400x) *Abies balsamea*, balsam fir.

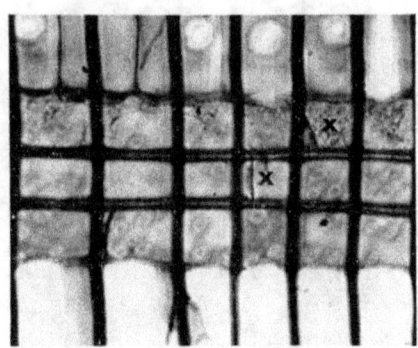

SF 13. Ray parenchyma end walls smooth (x in figure) (r, 400x) *Thuja plicata*, western red cedar.

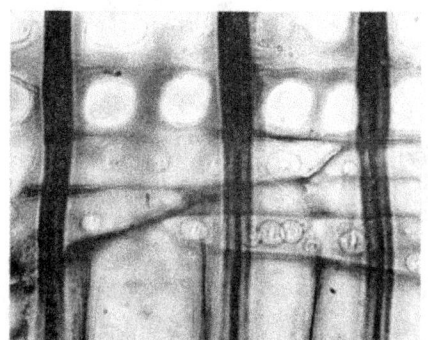

SF 14A. Fenestriform pitting (*r*, 400x) *Pinus strobus*, eastern white pine.

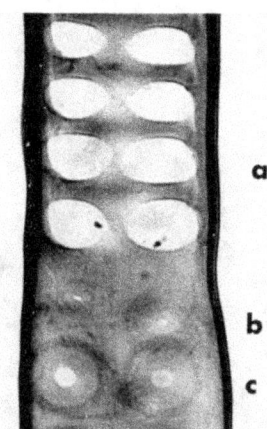

SF 14B. Fenestriform pitting (*pulp*, 400x) *Pinus strobus*, eastern white pine: (a) fenestriform pits leading to ray parenchyma, (b) small bordered pits leading to ray tracheid, (c) bordered pits leading to longitudinal tracheid.

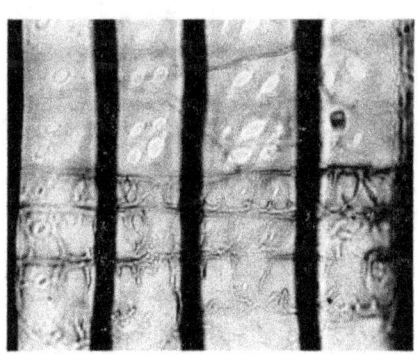

SF 15A. Pinoid pitting (*r*, 400x) *Pinus* spp., hard pine.

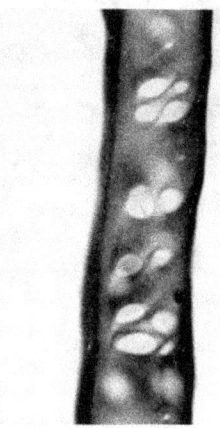

SF 15B. Pinoid pitting (*pulp*, 400x) *Pinus banksiana*, jack pine.

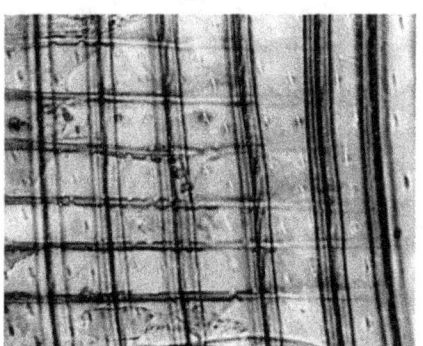

SF 16A. Piceoid pitting (*r*, 400x) *Picea rubens*, red spruce.

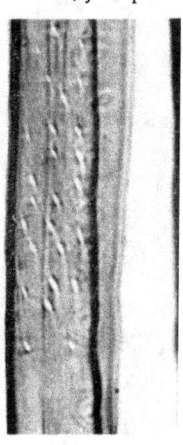

SF 16B. Piceoid pitting (*pulp*, 400x) *Picea rubens*, red spruce.

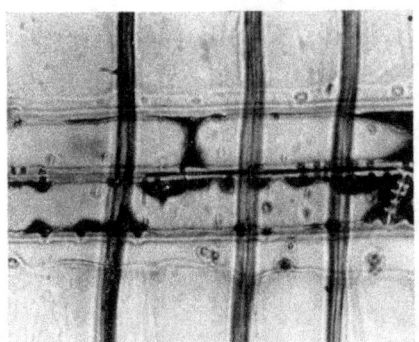

SF 17A. Cupressoid pitting (*r*, 400x) *Tsuga* spp., hemlock.

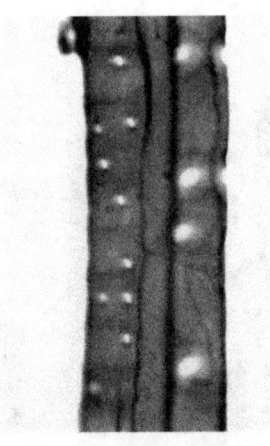

SF 17B. Cupressoid pitting (*pulp*, 400x) *Tsuga* spp., hemlock.

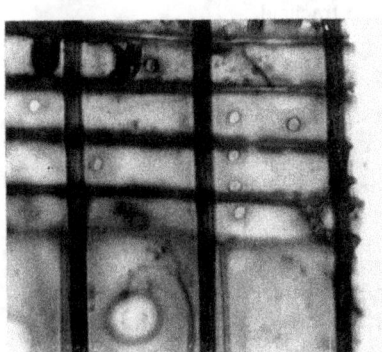

SF 18A. Taxodioid pitting (*r*, 400x) *Abies balsamea*, balsam fir.

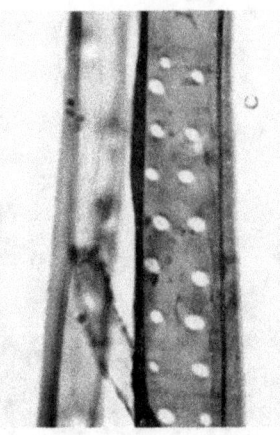

SF 18B. Taxodioid pitting (*pulp*, 400x) *Abies balsamea*, balsam fir.

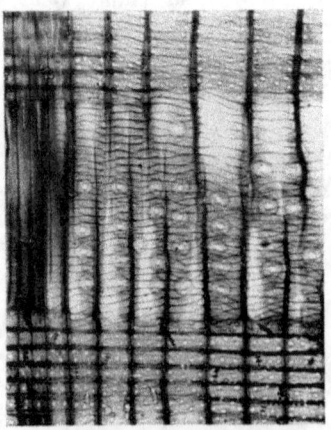

SF 19. Spiral thickening present (*r*, 125x) *Pseudotsuga menziesii*, Douglas fir.

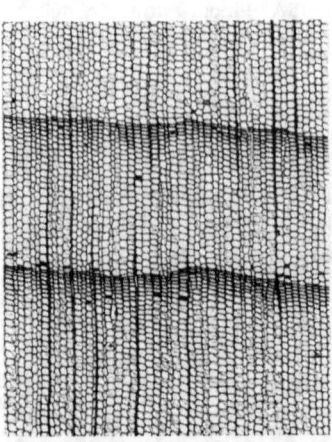

SF 20. Longitudinal parenchyma (black) in body of ring (*x*, 25x) *Chamaecyparis nootkatensis*, yellow cedar.

IDENTIFICATION KEY FOR COMMERCIAL PULPWOODS AND PULPS

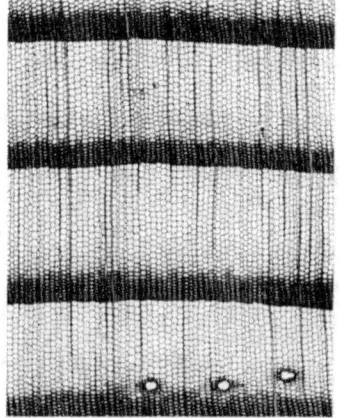

SF 21. Earlywood-latewood transition abrupt (*x*, 15x) *Psuedotsuga menziesii*, Douglas fir. Compare with SF 1 and 20 where transition is gradual.

SF 22. Dimpled grain (split tangential, 1.1x) *Pinus contorta*, lodgepole pine.

DIAGNOSTIC FEATURES FOR HARDWOODS

ABBREVIATIONS: *x*—cross-section; *r*—radial-section; *t*—tangential section; *pulp*—pulp fibre.
(See also illustrations of these features which are designated by the same numbers.)

I. Vessel Features

1. RING POROUS (*x*). Pores in the earlywood large, with an abrupt transition in size to those of latewood.
2. DIFFUSE POROUS (*x*). Pores fairly uniform in size, or very gradually reduced in size from earlywood to latewood.
3. ULMIFORM LATEWOOD (*x*). Pores in the latewood in wavy concentric bands paralleling the growth rings.
4. PERFORATION PLATES SCALARIFORM (*r, pulp*). End walls of vessel elements bear multiple bar-like structures.
5. PERFORATION PLATES SIMPLE (*r, pulp*). End walls of vessel elements with a single large opening devoid of bars.
6. SPIRAL THICKENING PRESENT (*r* or t, pulp*). Spring-like helix running the length of vessel elements, or less frequently, confined to the tips of the vessel elements.
7. SPIRAL THICKENING ABSENT (*r or t, pulp*). No evidence of spring-like helix over any portion of vessel element.*
8. INTERVESSEL PITS OPPOSITE (*t, pulp**). Pits in definite horizontal rows across a vessel element.
9. INTERVESSEL PITS ALTERNATE (*t, pulp**). Pits in diagonal rows across a vessel element.
10. INTERVESSEL PITS LINEAR (*t, pulp**). Pit apertures markedly elongated in the horizontal direction across a vessel element.

*Not illustrated.

II. Parenchyma Arrangement

Pattern of parenchyma tissue usually as seen with *hand lens magnification*. The parenchyma is generally visible as grey or whitish-coloured tissue against the darker background of wood fibres and vessels.

11. PARATRACHEAL (x). Parenchyma arranged in association with vessels, appearing as a white sheath surrounding pores.

12. APOTRACHEAL (x). Parenchyma arranged independently of vessels, appearing as several white lines within growth ring, and running in a direction parallel to the growth ring.

13. MARGINAL (x). Parenchyma confined to boundary of growth ring, appearing as a fine *white* line delimiting the growth ring.

III. Tracheids

14. VASCULAR (*pulp*). A cell resembling a latewood vessel, but without perforation plates at its ends.

15. VASICENTRIC (r, *pulp*). A cell intermediate in length between a fibre and a vessel element and exhibiting irregular outline and distinct pitting.

IV. Rays

16. HETEROCELLULAR (r). A ray composed of two distinct shapes of cells: procumbent, having a long axis in horizontal direction, and upright, having square, or long axis in vertical direction. Procumbent cells usually form the main body of ray, with upright cells restricted to margins.

17. HOMOCELLULAR (r). A ray composed entirely of procumbent (usual case) or upright cells.

18. EXCLUSIVELY UNISERIATE (t). All rays only one cell wide.

19. 1 TO 3 SERIATE (t). Rays varying from one to three cells wide when measured at their widest point (centre of ray).

20. 1 TO 5 SERIATE (t). Rays varying from one to five cells wide when measured at their widest point (centre of ray).*

21. 1 TO 8 SERIATE (t). Rays varying from one to eight cells wide when measured at their widest point (centre of ray).

22. 1 TO MORE THAN 12 SERIATE (t). Rays varying from one to more than twelve cells wide when measured at their widest point (centre of ray).

23. AGGREGATE (x). A group of narrow rays very closely spaced, appearing as a single wide, but inconspicuous; ray visible with hand lens magnification.

V. Colour

24. COLOURED HEARTWOOD. Colour other than cream or straw-brown evident.*

VI. Commercial Range (see map, page 22)

25. EASTERN CANADA–NORTHEASTERN UNITED STATES.
26. SOUTHEASTERN UNITED STATES.
27. WESTERN CANADA–WESTERN UNITED STATES.

*Not illustrated.

ILLUSTRATIONS OF HARDWOOD FEATURES

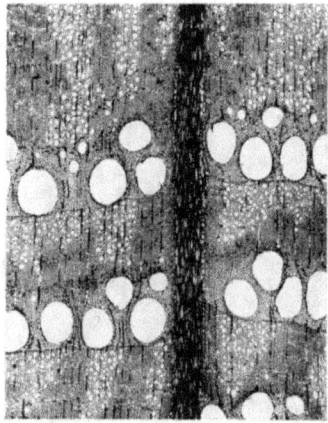

HF 1. Ring porous wood (*x*, 15x) *Quercus alba*, white oak.

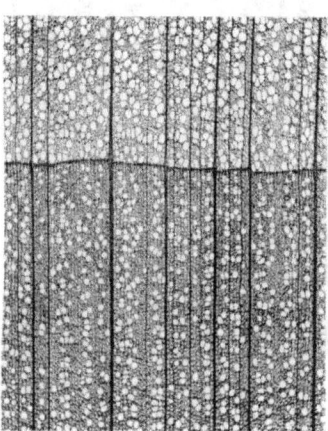

HF 2. Diffuse porous wood (*x*, 15x) *Tilia americana*, basswood.

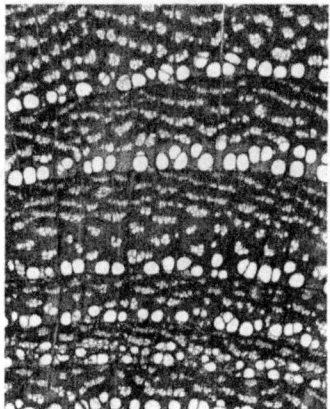

HF 3. Ulmiform latewood (*x*, 15x) *Ulmus americana*, white elm.

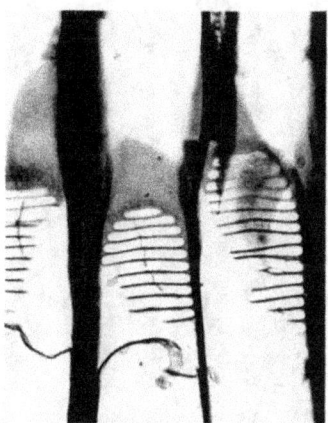

HF 4A. Perforation plates scalariform (*r*, 250x) *Liquidambar styraciflua*, red gum.

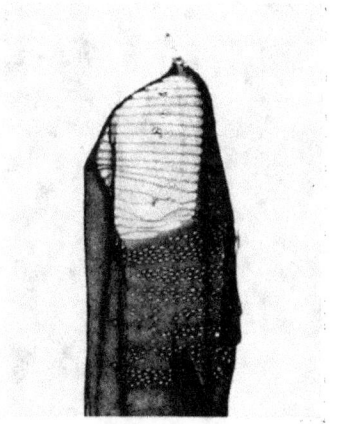

HF 4B. Perforation plates scalariform (*pulp*, 125x) *Alnus rubra*, red alder.

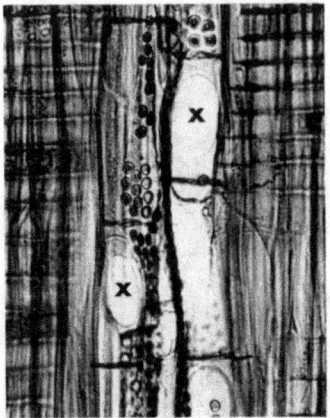

HF 5A. Perforation plates "x" simple (*r*, 250x) *Populus tremuloides*, trembling aspen.

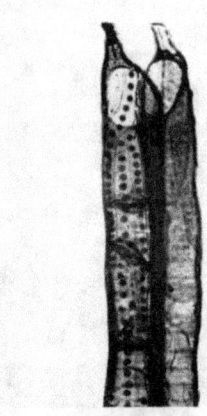

HF 5B. Perforation plates simple (*pulp*, latewood vessel elements, 125x) *Quercus alba*, white oak.

HF 6A. Spiral thickening present (*t*, 250x) *Tilia americana*, basswood.

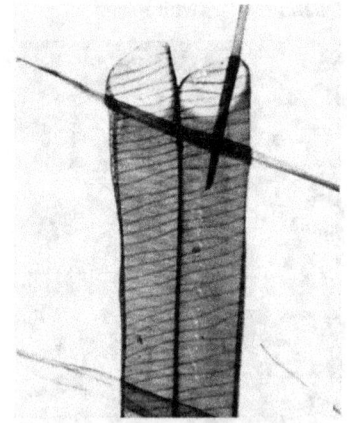

HF 6B. Spiral thickening present (*pulp*, vessel elements, 125x) *Tilia americana*, basswood.

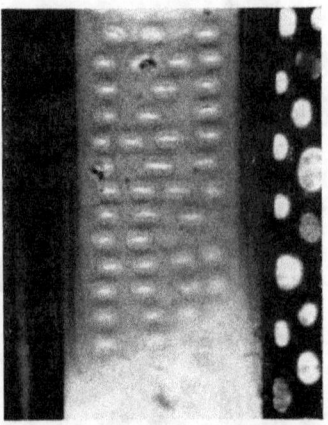

HF 8. Intervessel pits opposite (*t*, 400x) *Liriodendron tulipifera*, yellow poplar.

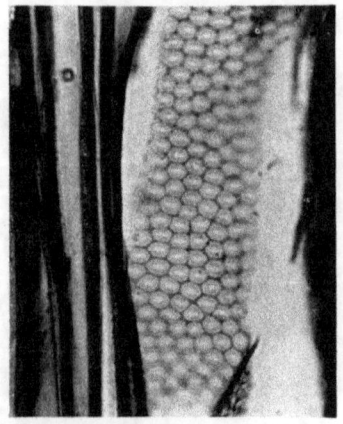

HF 9. Intervessel pits alternate (*t*, 400x) *Salix* spp., willow.

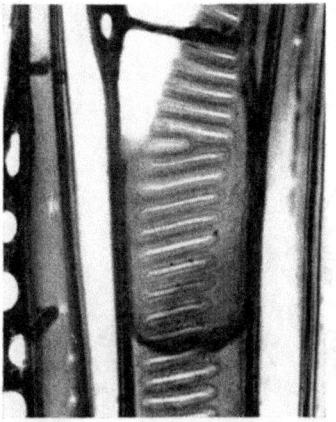

HF 10. Intervessel pits linear (*t*, 400x) *Liriodendron tulipifera*, yellow poplar.

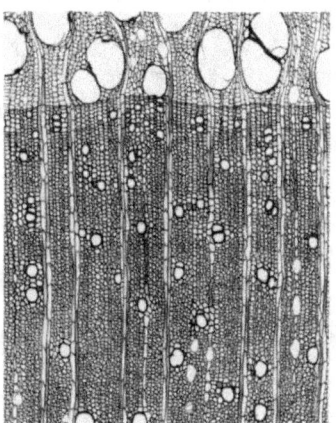

HF 11. Paratracheal parenchyma (*x*, 25x) *Fraxinus americana*, white ash.

HF 12. Apotracheal parenchyma (*x*, 25x) *Carya ovata*, shagbark hickory.

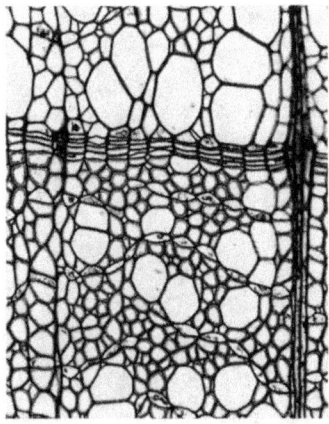

HF 13. Marginal parenchyma (*x*, 125x) *Tilia americana*, basswood.

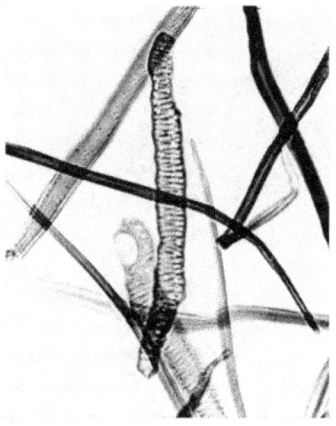

HF 14. Vascular tracheid in middle (*pulp*, 125x) *Ulmus americana*, white elm.

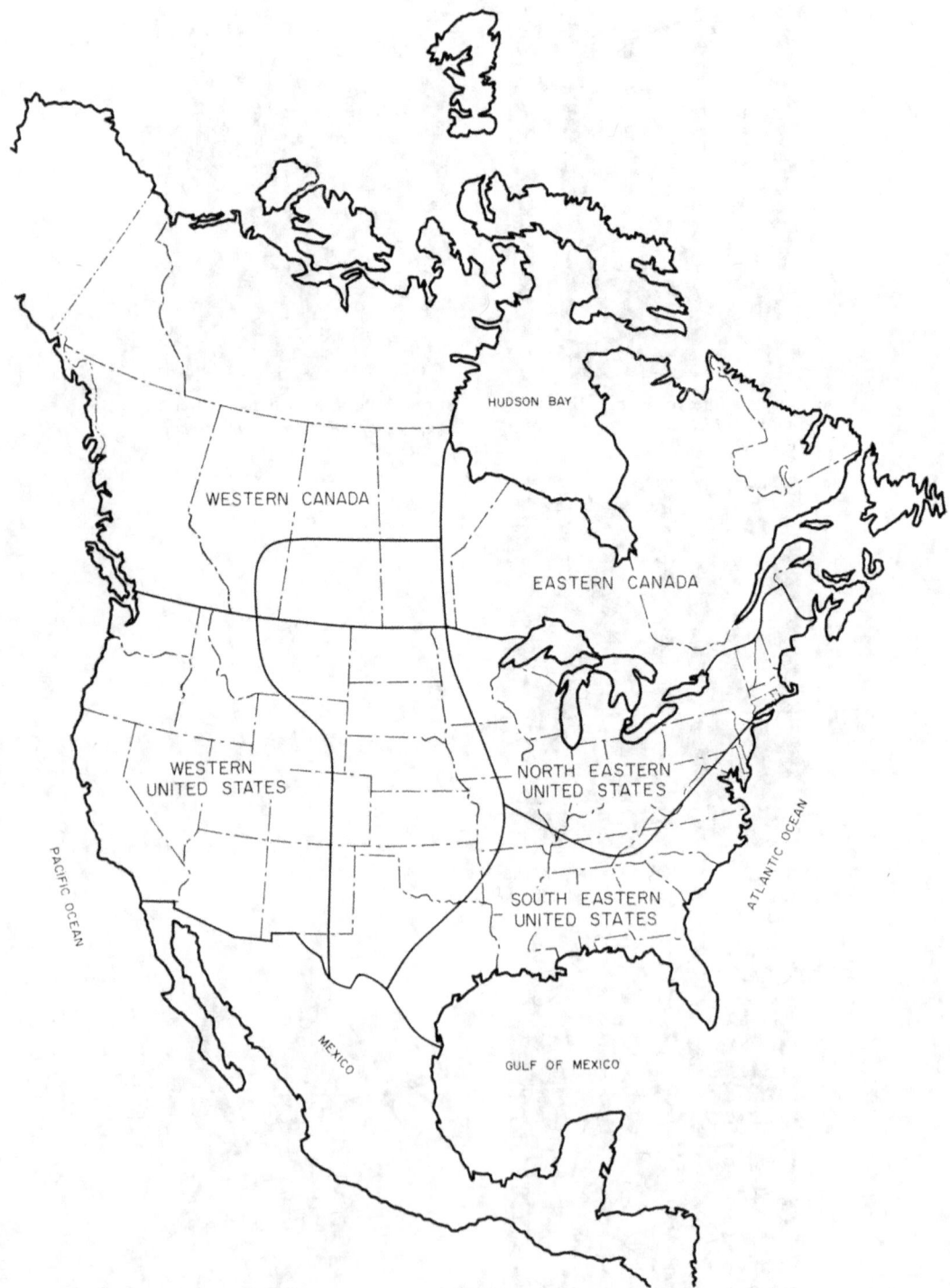

Commercial ranges of tree species included in key.

IDENTIFICATION KEY FOR COMMERCIAL PULPWOODS AND PULPS

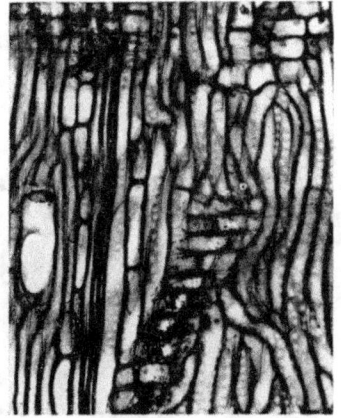

HF 15A. Vasicentric tracheids (r, 125x) *Quercus alba*, white oak.

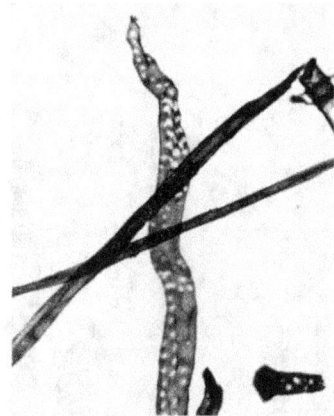

HF 15B. Vasicentric tracheid (*pulp*, 125x) *Quercus rubra*, red oak.

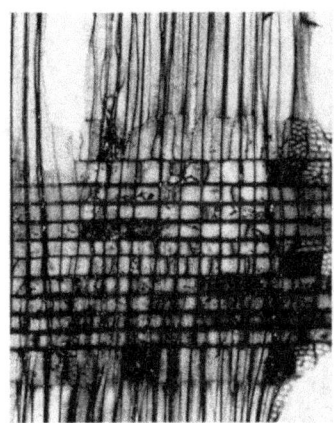

HF 16. Heterocellular ray (r, 125x) *Salix* spp., willow.

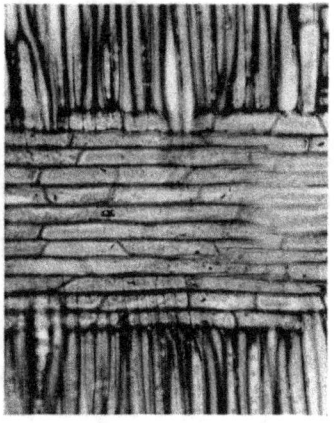

HE 17. Homocellular ray (r, 125x) *Fraxinus americana*, white ash.

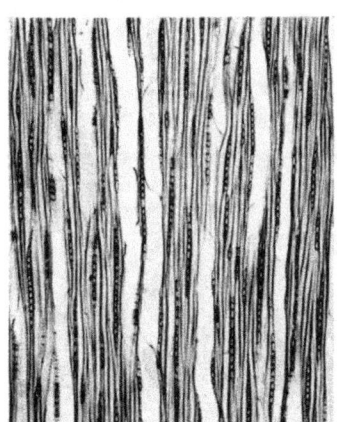

HF 18. Exclusively uniseriate rays (t, 60x) *Salix* spp., willow.

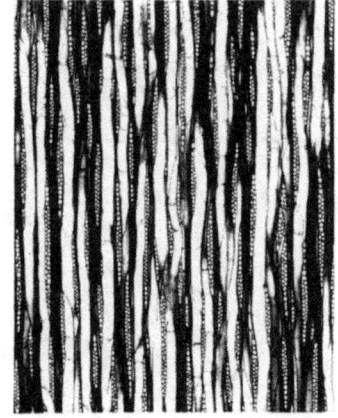

HF 19. Rays 1 to 3 seriate (t, 25x) *Liquidambar styraciflua*, red gum.

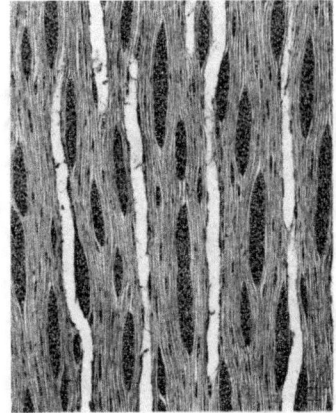

HF 21. Rays 1 to 8 seriate (*t*, 25x) *Acer saccharum*, sugar maple.

HF 22. Rays 1 to more than 12 seriate (*t*, 25x) *Quercus alba*, white oak.

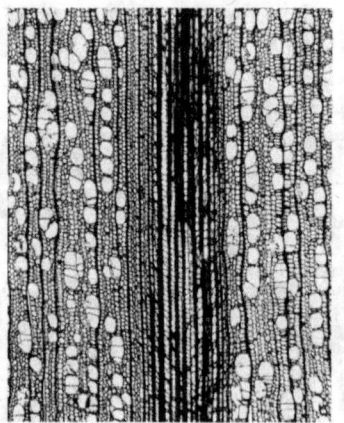

HF 23. Aggregate ray (*x*, 25x) *Alnus rubra*, red alder.

SOFTWOODS AND HARDWOODS

(The numbers accompanying each genus, species or group of species refer to the diagnostic features categorizing that particular group.)

Common names	Scientific names	Features
SOFTWOODS		
White pine (western)	*Pinus monticola*	25 ⎫
		1, 4, 7, 8, 9, 13, 14, ⎬
White pine (eastern)	*P. strobus*	23 ⎭
Red pine	*P. resinosa* ⎫	
	⎬	1, 4, 6, 9, 11, 13, 14, 21, 23
Scots pine	*P. sylvestris* ⎭	

IDENTIFICATION KEY FOR COMMERCIAL PULPWOODS AND PULPS

Common names	Scientific names	Features
Jack pine	P. banksiana	23
Lodgepole pine	P. contorta	1, 4, 6, 9, 11, 13, 15, 21, 22, 25
Ponderosa pine	P. ponderosa	1, 4, 8, 9, 11, 13, 15, 21, 22, 25
Southern pines: shortleaf,	P. echinata	
slash,	P. elliottii	
longleaf,	P. palustris	1, 4, 7, 9, 11, 13, 15, 21, 24
pitch,	P. rigida	
pond,	P. serotina	
loblolly,	P. taeda	
Sitka spruce	Picea sitchensis	1, 3, 5, 6, 9, 12, 16, 17, 22, 25
Engelmann spruce	P. engelmannii	25
White spruce	P. glauca	23
		1, 3, 5, 9, 12, 16, 17,
Black spruce	P. mariana	23
Red spruce	P. rubens	23
Eastern larch, tamarack	Larix laricina	23
		1, 3, 5, 9, 12, 16, 17, 21,
Western larch	L. occidentalis	25
Douglas fir	Pseudotsuga menziesii	1, 3, 5, 9, 12, 16, 17, 19, 21, 25
Hemlock (eastern)	Tsuga canadensis	23
		2, 9, 12, 16, 17,
Hemlock (western)	T. heterophylla	25
True firs, balsam	Abies amabilis	25
	A. balsamea	23
		2, 10, 12, 18,
	A. grandis	25
	A. lasiocarpa	25
Western red cedar	Thuja plicata	2, 10, 13, 18, 20, 21, 25
Yellow cedar	Chamaecyparis nootkatensis	2, 9, 10, 12, 17, 20, 25

HARDWOODS

Common names	Scientific names	Features
Oak	Quercus spp.	1, 5, 7, 8, 9, 11, 12, 15, 17, 22, 24, 25, 26
Ash	Fraxinus spp.	1, 5, 7, 8, 9, 11, 15, 17, 19, 25, 26
Elm	Ulmus spp.	1, 3, 5, 6, 7, 9, 14, 17, 21, 25, 26
Hickory	Carya spp.	1, 2, 5, 7, 9, 12, 16, 17, 20, 24, 25, 26
Willow	Salix spp.	2, 5, 7, 9, 13, 16, 18, 24, 25, 26
Poplar, aspen, cottonwood	Populus spp.	2, 5, 7, 9, 13, 17, 18, 25, 26, 27
Beech	Fagus grandifolia	2, 4, 5, 7, 9, 16, 17, 22, 24, 25, 26
Basswood	Tilia americana	2, 5, 6, 9, 13, 17, 20, 21, 25
Birch	Betula spp.	2, 4, 7, 9, 13, 17, 20, 24, 25
Red alder	Alnus rubra	2, 4, 7, 8, 9, 17, 18, 19, 23, 24, 27
Maple	Acer spp.	2, 5, 6, 9, 17, 20, 21, 24, 25, 26
Red gum, sweetgum	Liquidambar styraciflua	2, 4, 6, 7, 8, 10, 16, 17, 19, 24, 26
Black gum, tupelo	Nyssa spp.	2, 4, 6, 7, 8, 10, 16, 19, 26
Tulip tree, yellow poplar	Liriodendron tulipifera	2, 4, 7, 8, 10, 13, 16, 17, 19, 20, 24, 26

USE OF CARD SYSTEM

Application of these diagnostic features to an edge-notched card system is recommended to facilitate identification. Examples of card keys are illustrated on page 26. The features listed numerically for each group should be transferred to a card by notching the margin opposite each feature.

Softwood card key (white pine)

#	Field	Section
1	Long. normal present (x)	RESIN CANALS
2	Long. normal absent (x)	RESIN CANALS
3	Epithelium thick-walled (x,t)	RESIN CANALS
4	Epithelium thin-walled (x,t)	RESIN CANALS
5	Average diam. <70μ (x)	RESIN CANALS
6	Average diam. 90μ (x)	RESIN CANALS
7	Average diam. 120μ (x)	RESIN CANALS
8	Average diam. >120μ (x)	RESIN CANALS
9	Tracheids present (r,p)	RAYS
10	Tracheids absent (r,p)	RAYS
11	Tracheids dentate (r,p)	RAYS
12	Par. end walls nodular (r,p)	RAYS
13	Par. end walls smooth (r,p)	RAYS
15	Fenestriform (r,p)	CROSS FIELD PITS
16	Pinoid (r,p)	CROSS FIELD PITS
17	Piceoid (r,p)	CROSS FIELD PITS
18	Cupressoid (r,p)	CROSS FIELD PITS
19	Taxodioid (r,p)	CROSS FIELD PITS
20	Spiral thickening (r,p)	MISCELLANEOUS
21	Long. par. in body of ring (x)	MISCELLANEOUS
22	Early-latewood trans. abrupt (x)	MISCELLANEOUS
—	Dimpled grain (t)	MISCELLANEOUS
24	E. Canada – N.E.U.S.	RANGE
—	S.E.U.S.	RANGE
26	W. Canada – W.U.S.	RANGE

E. white pine, *Pinus strobus*
W. white pine, *Pinus monticola*

Example of a card key for a softwood (white pine).

Hardwood card key (oak)

#	Field	Section
1	Ring porous (x)	VESSELS
2	Diffuse porous (x)	VESSELS
3	Ulmiform latewood (x)	VESSELS
4	Perforations scalariform (r,p)	VESSELS
5	Perforations simple (r,p)	VESSELS
6	Spiral thicken. present (r,t,p)	VESSELS
7	Spiral thicken. absent (r,t,p)	VESSELS
8	Intervessel pits opposite (t,p)	VESSELS
9	Intervessel pits alternate (t,p)	VESSELS
10	Intervessel pits linear (t,p)	VESSELS
11	Paratracheal (x)	PARENCHYMA
12	Apotracheal (x)	PARENCHYMA
13	Marginal (x)	PARENCHYMA
14	Vascular (p)	TRACHEIDS
15	Vasicentric (r,p)	TRACHEIDS
16	Heterocellular (r)	RAYS
17	Homocellular (r)	RAYS
18	Exclusively uniseriate (t)	RAYS
19	1–3 seriate (t)	RAYS
20	1–5 seriate (t)	RAYS
21	1–8 seriate (t)	RAYS
22	1–>12 seriate (t)	RAYS
23	Aggregate (x)	RAYS
—	Coloured heartwood	COL.
—	E. Canada – N.E.U.S.	RANGE
—	S.E.U.S.	RANGE
27	W. Canada – W.U.S.	RANGE

Oak, *Quercus spp.*

Example of a card key for a hardwood (oak).

Identification of an unknown specimen can be made by first selecting the appropriate deck of cards (depending on whether the unknown is a hardwood or a softwood) and then inserting a wire into a feature recognized from examination of an unknown specimen. The card representing the unknown will drop from the deck, along with all others notched for the same diagnostic feature. Only these cards will be employed for succeeding steps, which are based on other features observed from study of the unknown specimen. Such repetition will eventually reduce the deck to the point where only a single card falls, thereby concluding the identification problem.

It is most important that the diagnostic features be employed in the *positive sense only*, meaning that *only those cards falling from the wire should be retained for further sorting*. If a given specimen lacks a particular feature, the implication must not be drawn that the feature is always absent. For example, longitudinal parenchyma may be present abundantly in some samples of western red cedar, and entirely absent in others. Because of its fairly frequent occurrence, this feature (20 on the key) should be notched on the card for western red cedar. If no parenchyma were noted in a specimen of this species and only those cards which did *not* fall from the deck were retained when selecting feature 20, western red cedar would have been discarded. The absence of a feature can be used in the positive sense only in those instances where it is specifically mentioned (e.g., feature 7 in the hardwood key, "spiral thickening absent").

Occasionally it will appear that two mutually exclusive alternatives have both been selected as characterizing a given species. For example, elm is listed as having spiral thickenings both present and absent. This is made necessary by the fact that its very large earlywood vessels often lack spiral thickenings, while the smaller, more abundant latewood elements regularly exhibit them.

A card-sorting key has several advantages over the more common dichotomous type. One can proceed to select those features he is most certain are present and often can complete identification in no more than three steps. For example, if ulmiform pores are noted in a hardwood, elm is immediately singled out; if window-like pits and smooth-walled ray tracheids are observed in a pure pulp, identification of white pine can be made in two steps. As one becomes more familiar with the characteristics of a particular wood and its pulp, it may be advantageous to include certain notes and sketches regarding identification on the reverse side of the individual cards for ready reference. The key may be enlarged at any time to accommodate additional species simply by adding another card, thereby avoiding the rewriting necessary when revising or increasing a dichotomous key.

IDENTIFICATION OF WOOD FIBRES

FIBRE IDENTIFICATION is a prerequisite for fibre analysis. It is made with the aid of a microscope and must be performed by an analyst with experience in fibre microscopy. The procedure for the examination of paper and paper fibres is outlined in special standards of the Canadian Pulp and Paper Association and the Technical Association of the Pulp and Paper Industry. Besides regular papers, there are tar, asphalt, rubber, and other specially treated papers. There are also highly coloured papers. The standards describe the ways in which samples of these greatly varying papers can be disintegrated, prepared as micro slides, and examined. Wood is the raw material most favoured by the paper industry, and softwoods are preferred because they are more suitable for paper products. Hardwoods are also used, but to a much lesser extent.

As is any plant tissue, wood is composed of cells which are arranged in a certain order, depending upon genus. Chemical pulping processes reduce wood to pulp, which is then composed of individual cells. Thus the anatomical relationship of these cells to each other, used in identification of wood samples, no longer exists and cannot be used as a diagnostic feature. The features destroyed by pulping are: resin canals (features 1 to 8), longitudinal parenchyma (feature 20), earlywood–latewood transition (feature 21), and dimpled grain (feature 22) in softwoods, and ulmiform latewood (feature 3), patterns of parachyma (features 11 to 13), and structure of rays and colour of heartwood (features 16 to 24) in hardwoods. Consequently, identification of wood used in production of a given pulp or paper product depends solely upon features found on individual pulp elements, usually referred to as "fibres."

Fibres may be identified either by the previous experience of the analyst or by a method of comparison. The use of photomicrographs and fibre atlases is of great assistance, but is no substitute for the use of authentic samples. In known samples an important advantage is gained in that it is possible to observe all the variations of diagnostic features in the same type of fibre, which is much more illuminating than a single feature shown in a micrograph. For the comparison method it is always useful to keep at hand a fibrary, a collection of authentic pulp samples. The fibre analyst should make frequent use of such samples and make himself familiar with the differences within each sample and between pulps of different woods, especially those with similar and nearly similar features.

The general appearance of some diagnostic features (e.g., piceoid, cupressoid, and taxodioid pits) depends upon the degree of pulping (high and low yields), bleaching, and mechanical treatments. Pulps of high yields have better preserved features than those of low yields. Highly beaten fibres are much mutilated. In the latter case, it is sometimes almost impossible to find a single cross-field area or vessel element which may provide a clue to identity of the pulp.

It is a common misbelief of laymen that fibre species can be identified by stains, but staining by itself will not identify species or genera. However, some staining reactions do help to distinguish between chemical softwood and hardwood fibres. Either Selleger's or Alexander's stain can be used, and Korn's modification of Alexander's stain seems to work even better. The staining reactions are not very reliable, as they depend upon pulping processes and pulping degrees. They are, however, often useful in quantitative estimation of mixtures. The softwood fibres develop various shades of red, while those of hardwoods develop shades of blue.[4]

Mechanical softwood pulps can also be distinguished from mechanical hardwood pulps by staining them with a 2 per cent aqueous solution of aniline sulphate (acidified by one drop of concentrated sulphuric acid in 50 ml). Aniline sulphate stains all groundwood fibres yellow. But if the fibres are counterstained (after blotting away the excess) with aqueous solution (1:5000) of methylene blue, the hardwoods take on a bluish-green colour, while the softwood fibres remain yellow.[4]

The constituents of *mechanical pulps* (also known as groundwood pulps) are more easily identified than those of chemical pulps. In the former, small fragments of wood with broken and torn ends are always present. Some of these fragments have undisturbed rays, resin canals, and vessels which enable the analyst to employ directly the identification key based on wood features.

SOFTWOOD PULPS

Softwood pulps are mainly composed of logitudinal tracheids which are called "fibres" for convenience. These slender fibres have tapering ends between 2 to 4 mm long and a length/width ratio of between 50 and 100. They make up about 90 per cent of the softwood volume. The remaining 10 per cent consists of ray cells and in a very few species (cedars) also some longitudinal parenchyma cells. Ray cells are small, often brick-shaped, and are largely lost during washing and screening of pulp. The remaining ones have diagnostic value (softwood features 9 to 13). Epithelial cells of resin canals are destroyed during cooking and are not detectable in chemical pulps.

Softwood fibres are subject to wide variation in appearance within a single species. Usually they are classified into earlywood and latewood fibres. The earlywood fibres with thin walls and large lumens occur in pulp in a collapsed state and under the microscope appear flat, wide, and ribbon-like. These fibres have many more bordered pits (doughnut-like) in their walls than latewood fibres. Double or even triple rows of bordered pits are often found on the wide fibres of earlywood, although single rows prevail on the narrower ones. Bordered pits are more numerous towards the ends of the fibres. As many variations of arrangement are observed in all softwoods, bordered pits are of no diagnostic value for distinguishing species or genera. The latewood fibres have thick walls and narrow lumens. They usually do not collapse during chemical pulping. Under the microscope they appear long, thick, and narrower than earlywood fibres. Latewood fibres have smaller and fewer bordered pits or none at all.

The longitudinal tracheids in softwoods are in contact with rays which cross them at irregular intervals. Those contact areas are known as cross-fields and also as ray contact areas. The cells of the rays are in contact with longitudinal tracheids through pits which are very different from intertracheid bordered pits. Cross-field pits are openings of various forms and sizes, depending upon species or genus, and are classified as fenestriform, pinoid, piceoid, cupressoid, and taxodioid (see softwood features 14 to 18).

When it comes to the identification of pulp fibres, these cross-field pits are virtually the only reliable features by which the various softwoods are distinguishable. However, the shape of the pits, as well as their arrangement in the cross-fields, does not always permit identification of species or genera but rather aids in the classification of softwoods into broad groups with similar cross-field features. Cross-field pitting in latewood tracheids is often atypical and therefore should be rejected for identification purposes.

By the type of pitting in earlywood tracheids, softwood pulps can be grouped as follows:

GROUP 1. The fibres of *eastern and western white pines* (SP 1 and 2) are not distinguishable without knowing the place of origin because they have similar fenestriform pits.

GROUP 2. The fenestriform pits of *red pine and Scots pine* (SP 3 and 4) are quite similar to those of the white pines, but there are some differences in size and arrangement. It would be difficult to distinguish the species in group 2 from those in group 1 when in mixtures, although their identification in pure pulps is aided by the presence of dentate ray tracheids in group 2.

GROUP 3. It is usually impossible to distinguish with certainty the fibres of *jack, lodgepole, ponderosa, and southern pines* (SP 5, 6, 7, 8) without knowing the place of origin, because all these species have pinoid pits. Typical pulps of southern pines have a much higher percentage of definite latewood fibres (approximately 50 per cent), and in general all their fibres are of greater diameters than those of other pines in this group. The relative weight factor of southern pine fibres is 1.55, while that of all other pines is only 0.9.[5] Because of this great difference in weight factors, the separation of southern pine fibres from those of other pines is of real importance in quantitative fibre analysis. Whether or not these fibres can be identified by the above mentioned features will depend upon the degree of fibre degradation resulting from chemical and mechanical processing, as well as on the analyst's familiarity with these fibres. Even then some degree of uncertainty will prevail.

Ponderosa pine can often be distinguished from jack or lodgepole pines, since most earlywood fibres of ponderosa pine have at least one cross-field with many typically arranged pinoid pits (SP 7). At least in the fibres of jack and lodgepole pines such an arrangement of cross-field pits is not usually observed.

GROUP 4. This group includes all the *spruces, larches, and hemlocks* (SP 9 to 17). Spruces are the main pulpwoods of Canada and their fibres are encountered in almost all paper products.

These three genera have ray crossing pits of *two* kinds. The ray parenchyma pits range from piceoid to cupressoid in form and preclude positive separation of genera. From one to a few small bordered pits (ray tracheid pits) also are present at most margins of ray crossings. These most closely resemble the cupressoid pits in appearance, but in heavily treated pulps their detection requires careful examination. The piceoid pitting predominates in typical earlywood fibres of spruces and larches, while cupressoid pitting is most abundant in eastern and western hemlocks. Because of the less extended pit apertures, hemlock fibres often are distinguishable from the other two genera. However, severe chemical, and especially mechanical, treatments may obscure this difference. In addition, mixtures of different genera obscure identification of any particular one that is present. On the other hand, in pure pulps the general appearance of the fibres, as influenced by width, length, fibre ends, arrangement of bordered and cross-field pits, may give an indication of the genus to an experienced fibre analyst.

GROUP 5. The tracheids *of Douglas fir* have the same cross-field pittings as those of the previous group and thus cannot be separated from other species on this basis. However, earlywood and most latewood tracheids exhibit spiral thickenings on the inner surface of the cell wall adjacent to the lumen. These spirals are easily observed under the microscope (SP 18) and occur only in Douglas fir.

GROUP 6. This group is composed of the *true firs, western red cedar, and yellow cedar* (SP 19, 20, 21). The fibres of the firs and western red cedar have mostly taxodioid cross-field pits and normally *do not* have small bordered pits at field margins. However, since yellow cedar rays can be composed either entirely of parenchymatous cells or of ray tracheids, the fibres of this species will also show cross-fields with only cupressoid pits or with only small bordered pits. It is extremely difficult, however, to recognize the difference between these two ray contact areas. Yellow cedar may be identified initially as a hemlock because of its closely related pitting, but careful examination will fail to reveal the marginal ray tracheid pits which are typical of the genera in group 4. Furthermore, the taxodioid pits of the true firs and western red cedar may integrate with the cupressoid pits of yellow cedar and hemlock, but the absence of marginal ray tracheid pits in the group 6 species will aid in their separation from hemlock.

Identification of the wood pulps of these three genera will depend upon (1) the experience in recognizing the typical pit arrangements in the cross-fields of earlywood fibres and (2) the degree of fibre degradation that has occurred during mill processing.

IDENTIFICATION OF SOFTWOOD PULPS

The following illustrations, of ray cross-field pits of earlywood tracheids, will aid in the identification of softwood pulps.

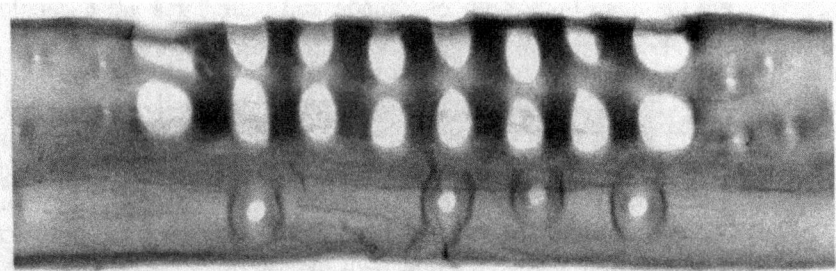

SP 1. *Pinus monticola*, western white pine (400x).

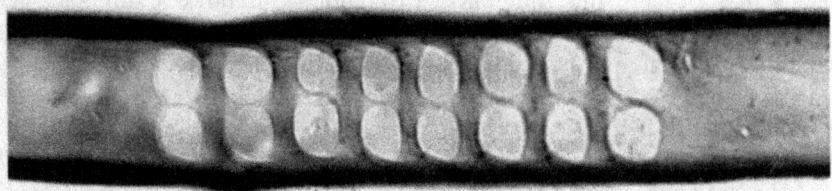

SP 2. *Pinus strobus*, eastern white pine (400x).

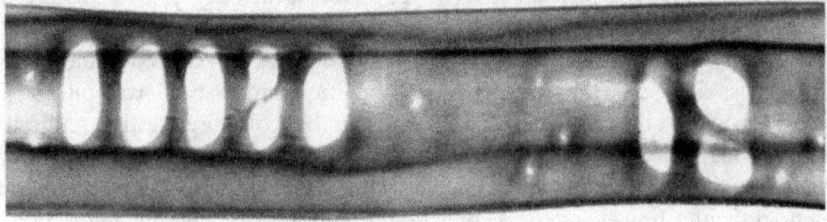

SP 3. *Pinus resinosa*, red pine (400x).

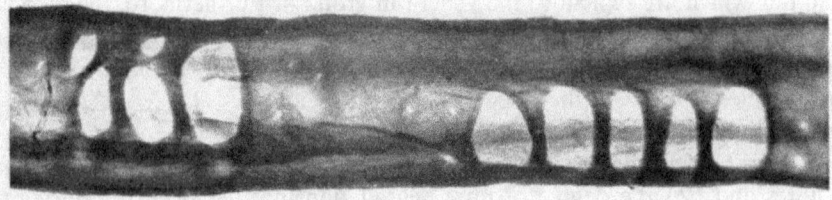

SP 4. *Pinus sylvestris*, Scots pine (400x).

IDENTIFICATION OF WOOD FIBRES 33

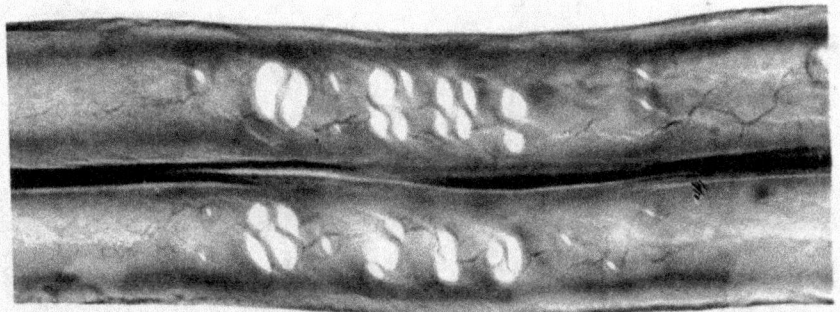

SP 5. *Pinus banksiana*, jack pine (400x).

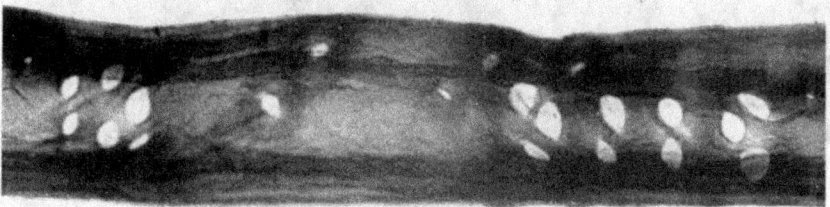

SP 6. *Pinus contorta*, lodgepole pine (400x).

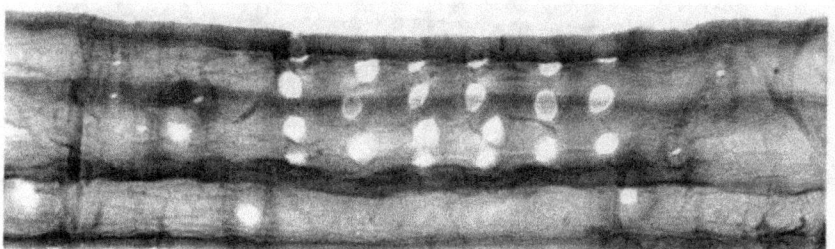

SP 7. *Pinus ponderosa*, ponderosa pine (400x).

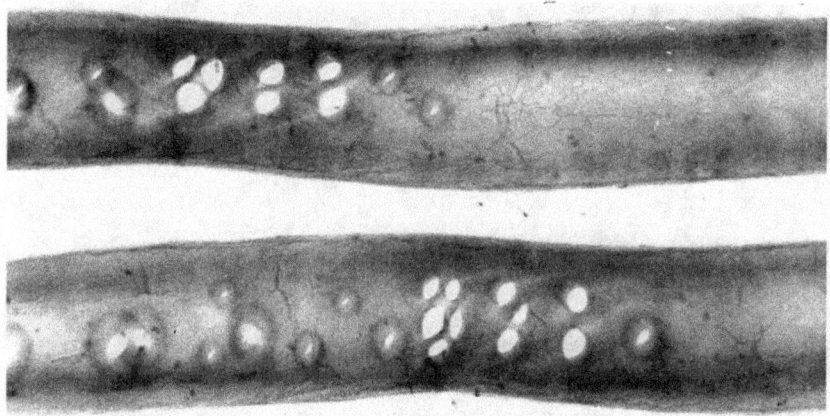

SP 8. *Pinus rigida*, pitch pine (400x).

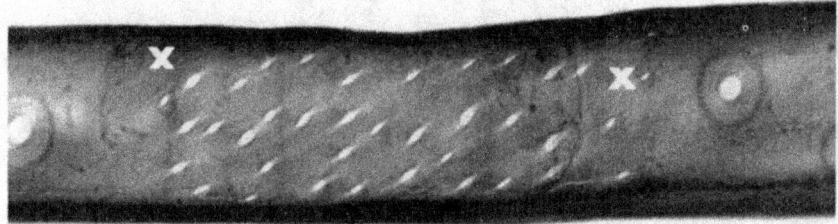

SP 9. *Picea sitchensis*, Sitka spruce (400x).

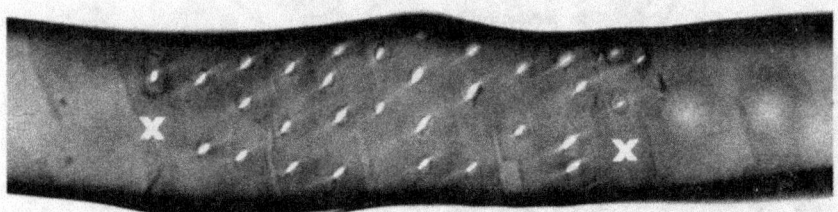

SP 10. *Picea engelmannii*, Engelmann spruce (400x).

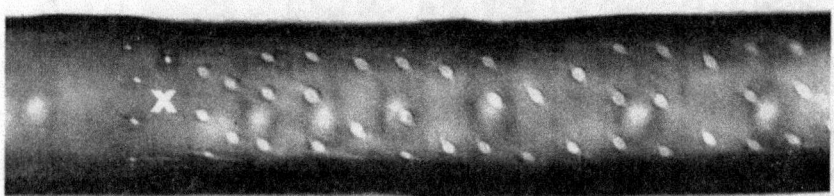

SP 11. *Picea glauca*, white spruce (400x).

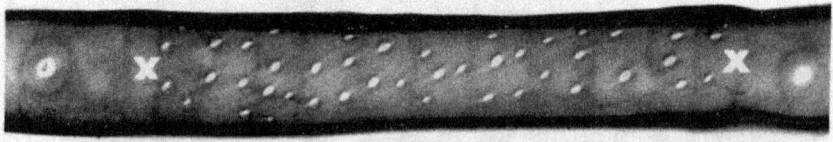

SP 12. *Picea mariana*, black spruce (400x).

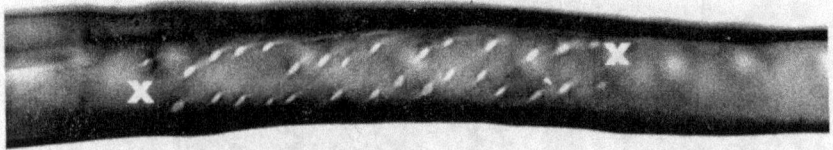

SP 13. *Picea rubens*, red spruce (400x).

NOTE: "X" marks small-bordered pits (ray tracheid pits) at the margins of cross-fields. See groups 4 and 6, p. 31.

SP 14. *Larix laricina*, eastern larch, tamarack (400x).

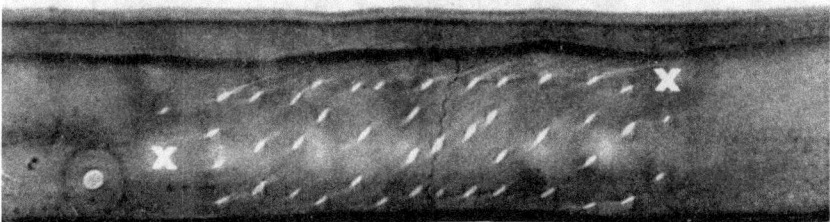

SP 15. *Larix occidentalis*, western larch (400x).

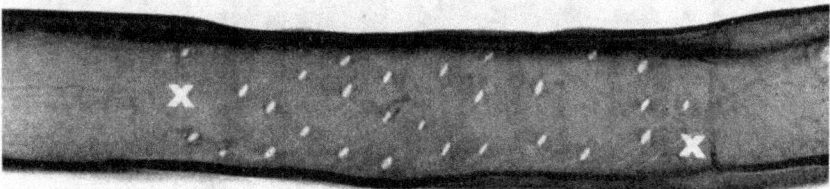

SP 16. *Tsuga canadensis*, eastern hemlock (400x).

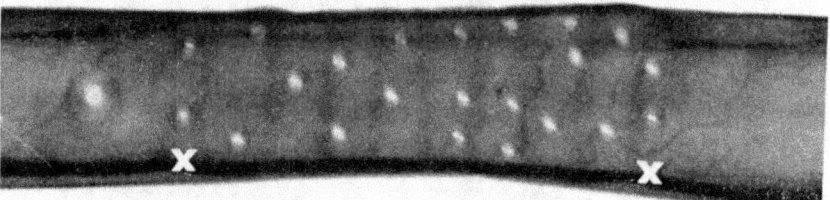

SP. 17. *Tsuga heterophylla*, western hemlock (400x).

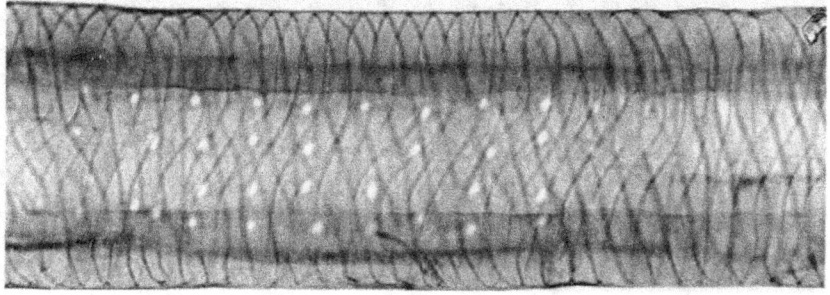

SP 18. *Pseudotsuga menziesii*, Douglas fir (400x).

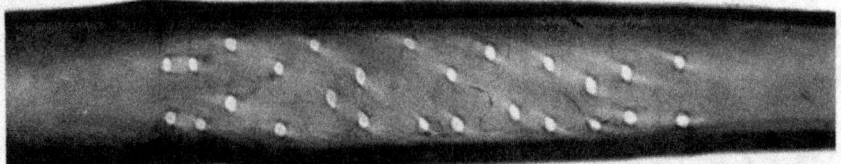

SP 19. *Abies balsamea*, balsam fir (400x). A representative of true firs.

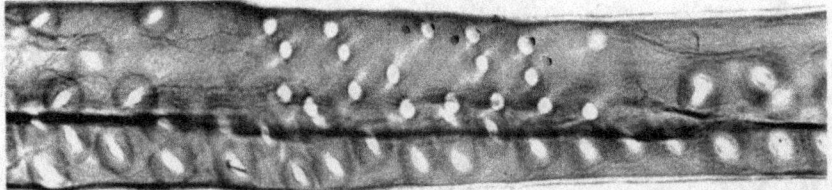

SP 20. *Thuja plicata*, western red cedar (400x).

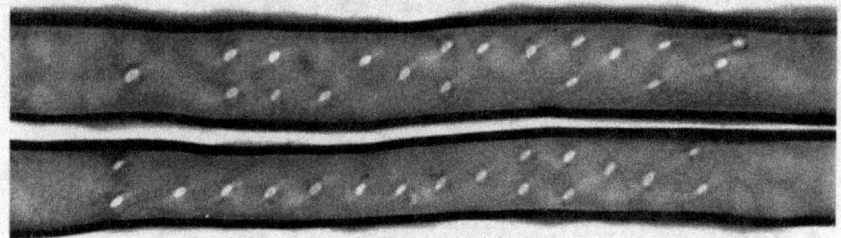

SP 21. *Chameacypatis nootkatensis*, yellow cedar (400x).

HARDWOOD PULPS

Because hardwoods are more complex in their anatomical structure than softwoods, their pulps have a greater diversity of cell types. Those usually observed are: vessel elements, fibre tracheids, libriform fibres, cells of longitudinal parenchyma, ray cells, and in some genera also vascular tracheids and vasicentric tracheids.

Vessel Elements

The pores visible in cross-section of a hardwood sample in reality are cross-sections of vessels. These vessels are longitudinal tubes in the structure of hardwoods and are composed of very many cells connected end to end. Those vessel components are known variously as vessel elements, vessel segments, or vessel members.

Vessel elements are the only type of cells showing constant structural variation between the different genera. Because of this, the identification of hardwood pulps is based solely on features observed in vessel elements. Although in any particular hardwood the features present remain constant, the size and shape of the vessel elements may vary greatly, depending upon whether an element has been derived from earlywood, latewood, or the transition between the two. The widest vessel elements are in the earlywood, while the smallest ones are in the latewood (HP 1, 2, 5). In diffuse porous woods the differences in size are much less pronounced than in ring porous woods. In beaten pulps, the vessel elements may be broken, and identification may have to be based on detected fragments of vessel elements.

Each vessel element has two ends which appear like oblique openings (HP 15). The openings in some genera bear multiple bar-like structures called *scalariform* perforation plates. Depending upon the genus, the size of the bars and the number of parallel openings between these bars vary greatly and are of diagnostic value (compare HP 16, 17, 19, 20, 21). When the end of vessel element appears as an unobstructed opening it is termed a *simple* perforation plate (HP 2, 4, and others).

Where a vessel in wood was originally in contact with adjacent vessels, large masses of bordered pits resulted. The shape, size, and arrangement of these intervessel pits are peculiar to each genus. The arrangement of these pits is termed "opposite" when the pits are in horizontal rows across a vessel element, "alternate" when the pits are in diagonal rows across a vessel element, and "linear" when the pit apertures are elongated in the horizontal direction across a vessel element (compare HF 8, 9, 10, pages 20 and 21).

Vessel elements also have pitted ray contact areas roughly equivalent to crossfields in softwoods. Here again the size, shape, and arrangement of the pits are distinctive features for each genus. Vessel elements possess other pits through which they are in contact with adjacent tracheids, fibres, and longitudinal parenchyma cells. Cross-fields, however, have an orderly horizontal arrangement of pits and are easily distinguished from these other pitted areas where the rows of pits are directed vertically.

Vessel elements of certain genera (elm, basswood, and maple) exhibit spiral thickenings running the length of the elements, while in another genus (red gum) the spirals are confined to the tapering tips of the cells (HP 19).

Other Pulp Elements

Vascular tracheids are similar to small vessel elements but differ in being imperforate at the ends. They are found in elm pulps (HF 14). Vasicentric tracheids are copiously covered with bordered pits, are of irregular outline, and vaguely resemble short, softwood tracheids. They are found in the pulps of oaks and ashes and are of diagnostic value (HP 3).

Fibre tracheids, together with the libriform fibres, make up the bulk (over 50 per cent by volume) of all hardwoods and are the main constituents of the hardwood pulps. Both are fibrous cells with thick walls and tapering and pointed ends. The difference between the two types is that the fibre tracheids possess small, bordered pits which are most abundant on the radial walls, while libriform fibres have simple pits, are smaller in diameter, and have narrower lumens. All stages of transition between the two types occur, and in many cases it is an arbitrary decision as to whether a fibre belongs to the first or second group. The two kinds of fibre are quite similar in all hardwoods, except for variation in average length, and have no practical value in the identification of a genus.

The most common type of longitudinal parenchyma found in hardwoods is the strand parenchyma. The cells of this parenchyma are relatively short, rectangular, or almost so, and often possess simple pits. In wood the cells are in strands along the grain. These cells can be frequently observed, though not always, in hardwood pulps which have not been too severely treated. However, cells of strand parenchyma have no diagnostic value for genus identification.

The short brick-shaped cells which come from the horizontal system, the rays, are very abundant in hardwood pulps. Many of these cells are lost during washing and screening of pulps. It should, however, be remembered that in most hardwoods the rays occupy a volume of between 10 and 20 per cent and in some oaks up to 30 per cent. Ray cells have no diagnostic value.

In spite of the fact that several types of wood cell have no diagnostic value, their recognition may still be of some interest to the fibre analyst.

IDENTIFICATION OF HARDWOOD PULPS

The following illustrations of vessel elements will aid in the identification of hardwood pulps.

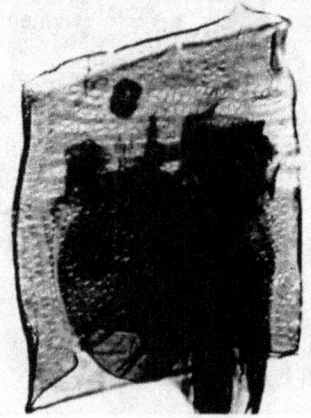

HP 1. Earlywood vessel element (125x) *Quercus alba*, white oak. Note tyloses in it.

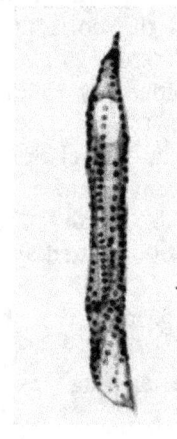

HP 2. Latewood vessel element (125x) *Quercus alba*, white oak.

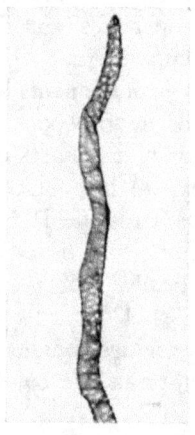

HP 3. Portion of a vasicentric tracheid (125x) typical for *Quercus* spp., oak.

HP 4. Earlywood vessel element (125x) *Quercus rubra*, red oak.

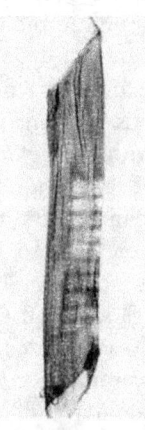

HP 5. Intermediate vessel element (125x) *Quercus rubra*, red oak.

HP 6. Earlywood vessel element (125x) *Fraxinus americana*, white ash.

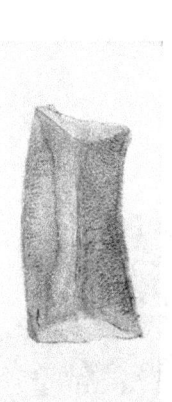
HP 7. Latewood vessel element (125x) *Fraxinus americana*, white ash.

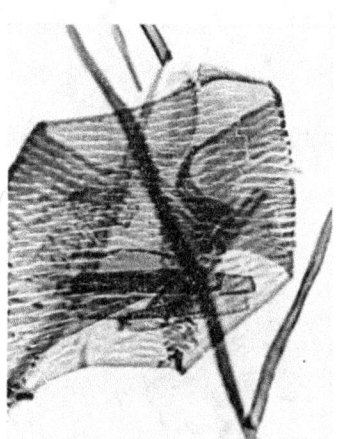
HP 8. Earlywood vessel element (125x) *Ulmus americana*, white elm.

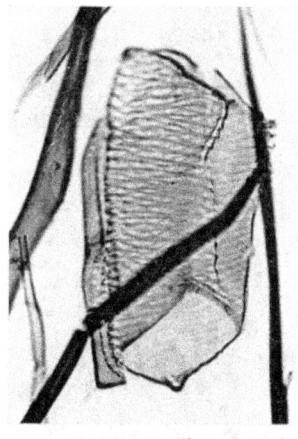
HP 9. Intermediate vessel element (125x) *Ulmus americana*, white elm.

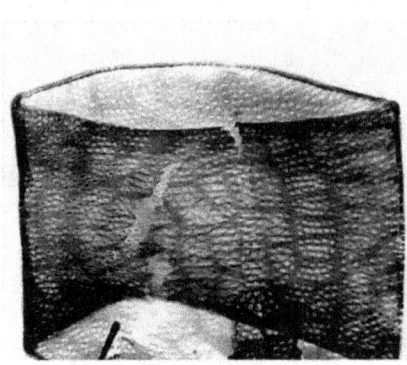
HP 10. Earlywood vessel element (125x) *Carya ovata*, Shagbark hickory.

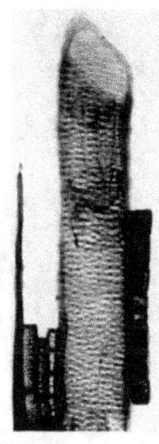
HP 11. Portion of a latewood vessel element (125x) *Carya ovata*, Shagbark hickory.

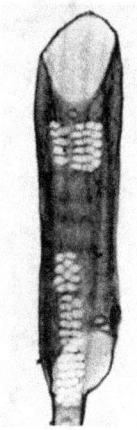
HP 12. Vessel element (125x) *Salix* spp., willow. Note ray contact areas parallel to long axis of vessel.

HP 13. Vessel element (125x) *Populus tremutoides*, trembling aspen.

HP 14. Vessel element (125x) *Fagus grandifolia*, beech. Note scattered arrangement of pits.

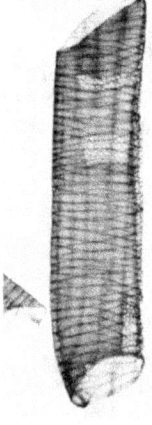
HP 15. Vessel element (125x) *Tilia americana*, basswood. Note spirals.

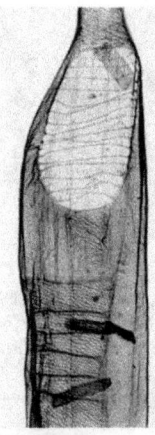

HP 16. Portion of a vessel element (125x) *Betula* spp., birch. Note numerous minute pits.

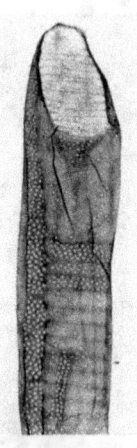

HP 17. Portion of a vessel element (125x) *Alnus rubra*, red alder.

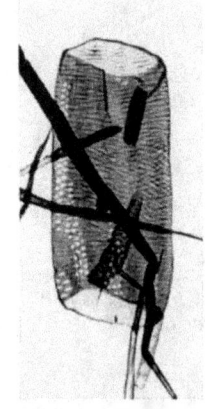

HP 18. Vessel element (125x) *Acer saccharum*, sugar maple. Note fine spirals.

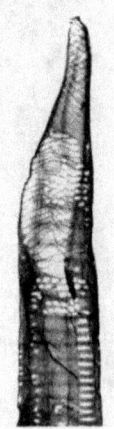

HP 19. Portion of a vessel element (125x) *Liquidambar styraciflua*, red gum. Note the spiral thickening confined to the tip.

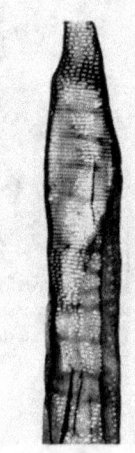

HP 20. Portion of a vessel element (125x) *Nyssa* spp., tupelo. Note many bars of perforation plate and opposite intervessel pits.

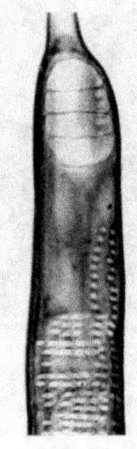

HP 21. Portion of a vessel element (125x) *Liriodendron tulipifera*, yellow poplar.

REFERENCES

1. SOKOLA, G. G. Bumazh. Prom. *36*, no. 2, pp. 14–15 (February, 1961) through Abstr. Bull. Inst. of Paper Chemistry *32*, pp. 210–211 (1961).
2. PANSHIN, A. J., *et al.* Textbook of Wood Technology, I, 2nd ed.; New York: McGraw-Hill, 1964. 643 pp.
3. CARPENTER, C. H., *et al.* Papermaking Fibres, Technical Publication no. 74, State University College of Forestry, Syracuse, N.Y., 1963. 77 plates.
4. ISENBERG, I. H. Pulp and Paper Microscopy, 3rd ed.; Appleton, Wisconsin: Institute of Paper Chemistry, 1958. 333 pp.
5. TAPPI standard T401m-60, TAPPI, 360 Lexington Ave., New York 17, N.Y. 10 pp.

PART TWO

IDENTIFICATION OF NONWOODY AND MAN-MADE FIBRES

I. Strelis

INTRODUCTION

MOST PAPER is made from wood fibres. But a large number of other fibres do find their way into paper production, including both vegetable and man-made textile fibres, other vegetable fibres such as grasses, and even animal and mineral fibres. Many specialty papers call for the use of fibres other than wood. In areas where wood supply is short, use is made of locally available raw materials, such as straw, bagasse, bamboo, etc. Consequently, to determine what material has been used in any particular paper, it is necessary to have means of identifying very specifically a wide variety of different fibres.

The various types of paper fibre are identified by their morphological features or staining reactions as revealed under the microscope. Those concerned with the identification of fibres have to acquire the basic knowledge of both the general appearance of various fibres and of their special characteristics.

The main purpose of this part is to deal with the microscopy of each type of fibre and to draw attention to the morphological features which are considered to be typical. To illustrate these features a visual aid is provided by the photomicrographs of carefully selected fibre samples which have been specially taken to show fibres whose features have not been too badly obscured by either mechanical or chemical treatment.

It is not easy to fully describe in words the features of a particular fibre-type, nor is it possible to display in detail, by means of a single photomicrograph, all the varying fibre forms and other cellular elements that occur. Photomicrographs are very useful in identification of fibres, but they are no substitute for experience and the use of authentic samples. If possible a collection of authentic samples, a library, should be kept for reference purposes. Observations and tests carried out on fibres of known origin are much more illuminating than either verbal descriptions or the exhibiting of a few fibres in a micrograph.

There are many man-made fibres and many more natural fibres. This part provides descriptions and photomicrographs of the more common ones which may be encountered in paper and paper products.

FIBRE IDENTIFICATION

The form and features of most paper fibres are not visible to the naked eye: they can be seen properly only under the microscope. Because of this, all existing methods on the fibre analysis of paper have been written for microscopic examination.

Fibre identification is a prerequisite for fibre analysis. The procedures for the examination of paper and paper fibres are outlined in special standards of the Canadian Pulp and Paper Association and the Technical Association of the Pulp and Paper Industry. Besides regular papers there are tar, asphalt, rubber, and other specially treated papers. There are also highly coloured papers. The standards describe the ways in which samples of these greatly varying papers can be disintegrated, prepared as micro slides, and examined.

This part is concerned only with the identification of fibres other than wood, including natural fibres, of vegetable, animal, or mineral origin, and manufactured fibres, of synthetic or regenerated origin.

NATURAL FIBRES

Natural fibres of vegetable origin are the most important group. They are the fibres most frequently found in paper furnishes. This group consists of:

(a) Seed hairs.

(b) Bast fibres—fibres derived from the bark of dicotyledons, which include herbaceous plants, shrubs, and trees.

(c) Leaf fibres—fibres derived from the vascular bundles of very long leaves of some monocotyledons. Leaf fibres are also known as "hard" fibres because they are more lignified than bast fibres.

(d) Grass fibres—these are another group of monocotyledonous fibres where the entire stem together with the leaves are pulped and used in papermaking. Such papers are composed not only of fibres, but of other cellular elements as well.

Since the chemical composition of various plant fibres is quite similar, staining and solubility tests do not reveal any significant differences between them. In addition it should be emphasized that the same type of fibre can have different colours with the same fibre stain, depending on whether it is in the raw, cooked, bleached, or unbleached state. Consequently, identification of vegetable fibres is based almost solely on their morphology.

Fibres mounted in water, glycerol, or a water-glycerol mixture reveal most structural details, while fibres mounted in Herzberg, C, Sutermeister, and other stains reveal the chemical conditions as well as some additional physical features. Although stains do not by themselves identify species, they do help in identification. For example, when a mixture of unbleached chemical wood and rag fibres is stained with Herzberg stain, they stain blue and red respectively. Such differential staining is useful because it divides the fibres into two groups. However, the wood may be one of many species and the rag might be of cotton, linen, or both, and it is still necessary, therefore, to identify the fibres specifically by morphological features.

To distinguish genera or species, it is important to take into account the following chracteristics, some of which result in quick and positive identification:

(*a*) The presence or absence of cells other than fibres: parenchyma, epidermal, vessel segments, and rings from annular vessels.

(*b*) The general shape of fibre: ribbon-like, cylindrical, of even or uneven width.

(*c*) Length and width.

(*d*) The particular shape of fibre ends: slender, pointed, blunt, thickened, or forked.

(*e*) The lumen: prominent, indistinguishable, wide, narrow, continuous, of even or uneven width, interrupted at intervals.

(*f*) The presence or absence of surface cross-markings, striations, pitting, spiral thickenings, or constrictions.

Identification of fibres is often facilitated by the use of polarized light. The various cross-markings, dislocations, and spirals appear bright, while the rest of the fibre is in the extinction position.

Difficulties in identification can be experienced because of similarities in morphology. Also, the chemical and mechanical treatments to which the fibres have been exposed can obscure, or even obliterate, features required for identification. In addition mixtures of two, three, or more species can easily obscure identification of any one of the component species.

Further difficulties may also arise from lack of experience. For example, flax and hemp are the most common bast fibres. They are quite similar in their general appearance and dimensions. A comparison of authentic samples reveals that their cross-markings are different in appearance; the lumen of hemp is wider than that of flax, the fibre ends of flax are slenderer and finer pointed than those of hemp. The use, however, of concepts like "different," "wider," and "finer" does not adequately define the differences observed. They serve only to draw the attention of the analyst who then compares authentic samples of the two fibres in order to observe what is meant by these terms.

The pulps of cereal straws, sugar cane bagasse, cornstalks, esparto, bamboo, and other grasses contain fibres as well as a great variety of other cells. This wide variety of cells is readily detected. However, it is often difficult to identify the species because their cellular elements are of similar appearance. Some species are distinguishable on the basis of having either large or small parenchyma cells or vessel segments. Others, like the cereal straws, all of which have cells and vessels of medium size, are practically indistinguishable from one another.

Wool fibres and animal hairs, because of their irregular horny scales, are readily distinguishable from other paper fibres. Identification of the animal origin of these fibres depends on slight variations of general features, further description of which is beyond the scope of this book.

MAN-MADE FIBRES

Paper has a multitude of performance requirements and to satisfy these has been produced commercially in recent years from blends of natural and man-made fibres or even from man-made fibres alone. The use of these fibres is relatively new, but the number of varieties used is steadily increasing, and the ability to identify them becomes more and more important.

A number of features are indicative of man-made fibres: because they are produced in continuous filaments, they exhibit two cut ends, they show evenness of diameter, and structural features are repeated along the fibre length. Within this group there are fibres with such similar surface characteristics that care has to be taken in their identification. Microscopic surface features alone should not be relied upon. Because manufactured fibres differ in chemical composition they are soluble in different solvents, and as a result *solubility tests* play an important role in their identification. Microsolubility or selective staining tests should always be used to confirm any conclusions arrived at from observing the morphology.

Solubility tests may be carried out in a dish or a watchglass but will yield more information if conducted under the microscope. In the case of fibre mixtures, tests should always be performed under the microscope in order to see the different reactions that occur. To carry out such tests, the dry fibres are placed on a slide and covered with a coverglass; the micro slide is then put under the microscope and brought into focus. The solvent is added with a dropper to the edge of the coverglass, and the reaction is observed. (See solubility tables Appendix I and II.)

Another important means of identification of man-made fibres is *selective staining*. A number of selective stains—mixtures of dyes which have different affinities for different fibres—have been developed by dye manufacturers and are used according to their specific recommendations. Samples for selective staining should be undyed, wettable, and free from extraneous matter such as fillers and sizes. In the case of coloured paper, the sample has to be stripped of all colour, and the stripping agent has to be completely washed out before the staining test is made. The fibres are identified by the resulting colours. (See Appendix III.)

Many fibre types possess distinctive cross-sections. Because of this, *indentification by cross-sections* is heavily relied upon with textiles. Books on the identification of textile fibres will usually give photomicrographs of the cross-sections as well as the longitudinal view of individual fibre types. This is because textile fibres are parallel

FIBRE IDENTIFICATION 47

oriented in threads and yarns where cross-sectioning presents no problem. On the other hand, paper fibres are oriented at random, and a section through paper will reveal few, if any, true fibre cross-sections. Consequently, cross-sections are not used in the identification of paper fibres.

IDENTIFICATION KEY

An attempt has been made to prepare a useful key for identification of the fibres described in this part (see Appendix IV). The key is based largely on microscopic morphological features, but for man-made fibres the results of solubility and staining tests have also been introduced. The key summarizes the more important features for distinguishing different fibres. Where there are definite diagnostic features it should lead directly to positive identification of a specific fibre type. In other cases this key will lead only to groups of fibre types having similar features. Final identification here will require comparison of the unknown sample with fuller descriptions, photomicrographs, and any available authentic samples. Whether or not given fibres can be identified will depend upon (1) the degree of their similarity to some other fibres, (2) the degree of their degradation due to chemical and mechanical treatments, and (3) the analyst's familiarity with these fibres.

SEED HAIRS: FIBRES

COTTON
(*Gossypium* spp.)

COTTON FIBRES are seed hairs that grow on the seeds of several species of the genus *Gossypium*. These plants are annual shrubs that reach the height of 3 to 5 feet. Cotton is cultivated in many countries of which the United States is the largest producer. Most American cotton varieties belong to the species *G. hirsutum* and are known as Upland Cottons. The cotton which is used for papermaking usually begins as a textile fibre. It finds its way into paper as either textile trimmings or as old rags.

Lint and Linters
The fibres attached to a single mature cotton seed vary considerably in length. These fibres fall into two groups: longer fibres, known as "lint," and shorter fibres called "linters." The latter remain on the seed as an undercoat of fuzz after the first ginning has removed the lint fibres. The microscopic appearance of linters is similar in many respects to that of lint cotton. They are, however, shorter and more cylindrical. They have thicker walls, narrower lumens, and are darker in colour.[2]

The outside of a raw cotton fibre is covered with a thin cuticle which consists of wax and pectin materials. The presence of wax can be demonstrated with Sudan III dye solution, which has an affinity for waxes and stains the cuticle an indefinite orange. Bleached cotton does not give any colour reaction with Sudan III because its wax cuticle has been removed.[4] The Sudan III dye solution is prepared by saturating a mixture of three parts of alcohol and one part of water with the dye and adding two parts of this solution to one part of glycerine.[2]

Because of typical characteristics, cotton is readily distinguishable from all other fibres (Figs. 1 to 3). It is a long, single cell fibre, ribbon-like in shape, with a large continuous lumen throughout its length. It is slightly thickened at the edges. Cotton has convolutions at irregular intervals. These convolutions show reversals of direction. They are present in the same fibre in both clockwise and anti-clockwise directions in about equal proportions.[5] Each fibre has a tapering tip and a broken base where the fibre was parted from the seed. Immature fibres have practically no convolutions.

The lumen is often partly filled with residual protoplasmic material. In cross-section mature fibres are elliptical or circular in shape, whereas immature fibres are often U-shaped. Longitudinal and spiral striations in the cell walls are common. These can be better observed in swollen fibres under polarized light.[1]

The cotton fibre has a primary wall which is composed of tiny fibrils. This wall resists the action of acids and cellulose solvents. The secondary wall is composed of successive layers of cellulose which represent daily growth rings. These layers are visible in cross-sections of fibres which have been swollen in cuprammonium hydroxide solution. Each daily ring consists of two layers, one compact and one porous.[1]

Mercerized Cotton

To mercerize is to treat yarn or cloth, usually kept under tension, with sodium hydroxide solution. Cotton is mercerized to improve its lustrous appearance and dyeing qualities. During mercerization, the fibres swell and lose most of their convolutions. However, rare convolutions or indications of convolutions by which the fibres can still be identified as cotton can always be found.

Mercerized cotton fibres (Figs. 4 and 5) are almost cylindrical and have smooth surfaces. They are also more lustrous than untreated cotton. The lumen appears as a fine line. Cotton also dyes more rapidly and to a darker shade after mercerization. Dyeing tests can be performed by placing both mercerized and unmercerized samples in the same bath with a direct dye.[1]

Both forms of cotton fibre stain red with Herzberg stain. Fibres of different species and varieties of cotton range in length from 1,000 to 4,000 times their widths.[7]

FIBRE LENGTH: 10 to 40 mm, average 18 mm.
FIBRE WIDTH: 12 to 38 microns, average 20 microns.

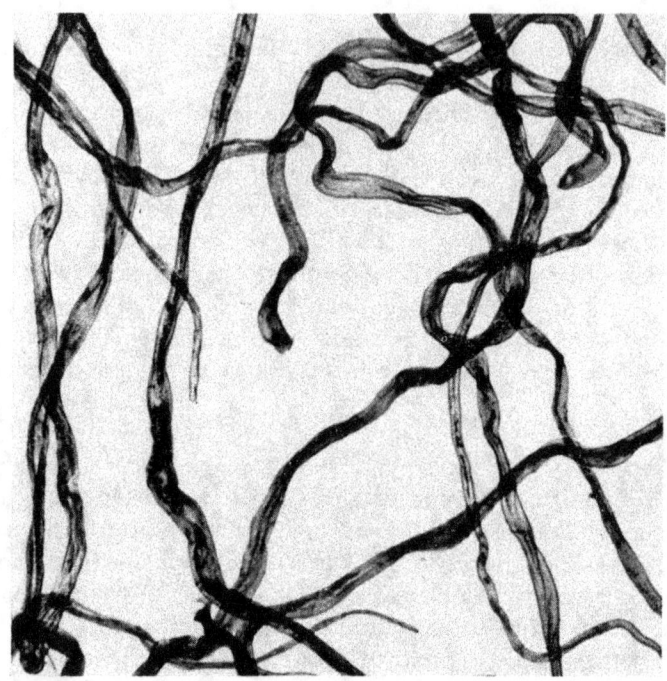

FIG. 1. Cotton fibres with protoplasmic residue in the lumens (150x).

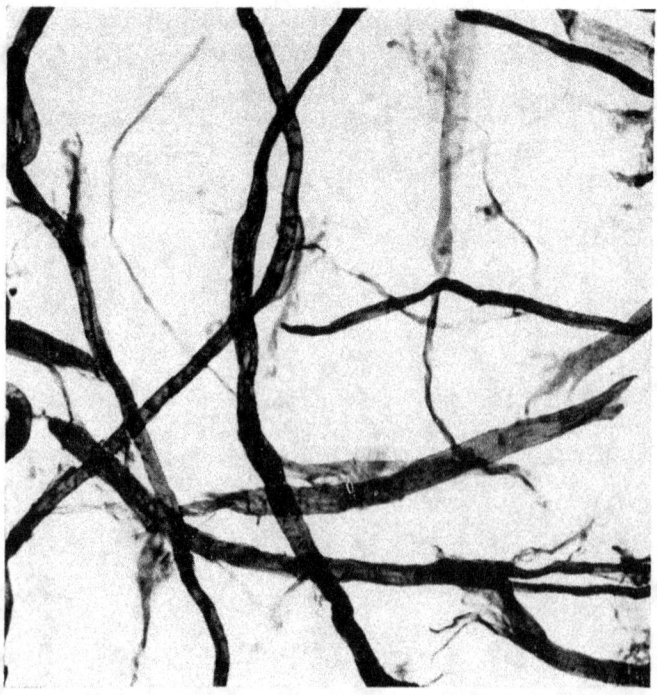

FIG. 2. Beaten cotton fibres (150x).

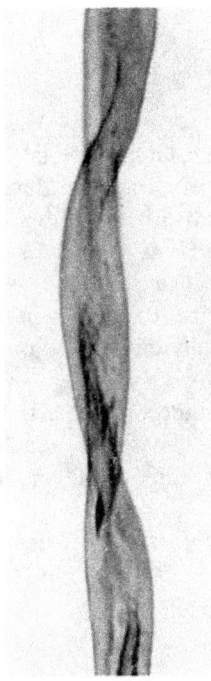

FIG. 3. Mid-portion of a mature cotton fibre (500x).

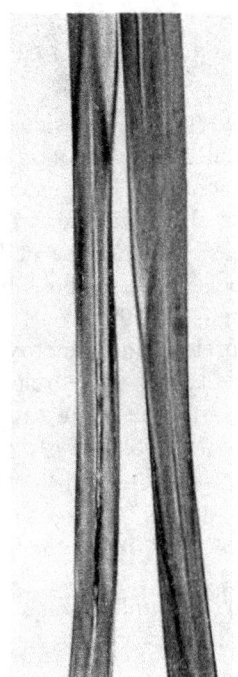

FIG. 4. Mid-portions of mercerized cotton fibres (500x).

FIG. 5. Mercerized cotton fibres (150x).

JAVA KAPOK
(Ceiba petandra)

The kapok fibre is a seed hair. It is obtained from the inside of the seed husk of the kapok tree grown in Indonesia. Kapok is mainly used as a stuffing material for upholstery and the manufacture of life jackets or other forms of protective device for lifesaving at sea. It has also been used as a substitute for absorbent cotton.[1,4]

Java kapok fibres are easily identifiable under a microscope (Figs. 6 and 7). A fibre consists of a single cell with a bulbous base. It is smooth, transparent, cylindrical, having a wide lumen, and thin walls and is frequently bent over on itself. The fibre surface is structureless, and the cross-section is generally circular or oval. In general, kapok fibres resemble structureless rods or tubes.[1]

When kapok fibres are immersed in a drop of water, the lumen only partially fills and many cylindrical air bubbles are formed in the lumen. These air bubbles are very noticeable under the microscope, are very typical of kapok, and are of unique value in identification.

The walls of kapok fibres are highly lignified and hence give a yellow-brown colour with Herzberg stain.

FIBRE LENGTH:[8]   8 to 30 mm, average 19 mm.

FIBRE WIDTH: 10 to 30 microns, average 19 microns.

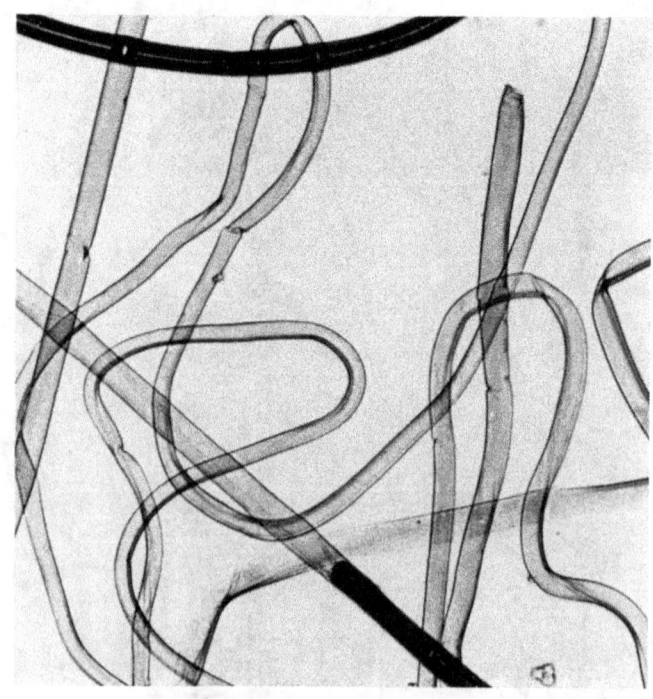

FIG. 6. Java kapok fibres (150x).

SEED HAIRS: FIBRES

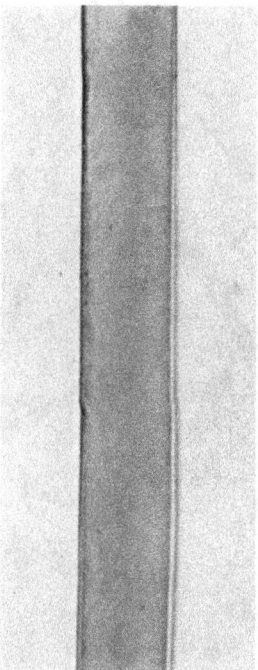

FIG. 7. Java kapok fibre (500x).

COIR
(*Cocos nucifera*)

Coir fibres are seed hairs which are obtained in brownish bundles from the husk that surrounds the nut of the cocnut plant. The main countries of coir production are India and Ceylon.

In spite of the relative shortness of the fibres, they are used to make cordage, coarse cloths, fish nets, brushes, and door mats.[1] Coir fibres are extremely resistant to wear and decay.[4]

Coir fibres are easily identifiable. They are rather short and wide with well-defined and broad lumens (Figs. 8 and 9). The fibre wall is thick, but rather irregularly so, giving the lumen an irregular intended outline. In most cases the fibre ends terminate abruptly and are blunt. Many pits, with diagonal striations between them, are observed in the fibre walls. Coir fibres are highly lignified.[1]

FIBRE LENGTH:[8] 0.3 to 1.0 mm, average 0.7 mm.
FIBRE WIDTH: 12 to 24 microns, average 20 microns.

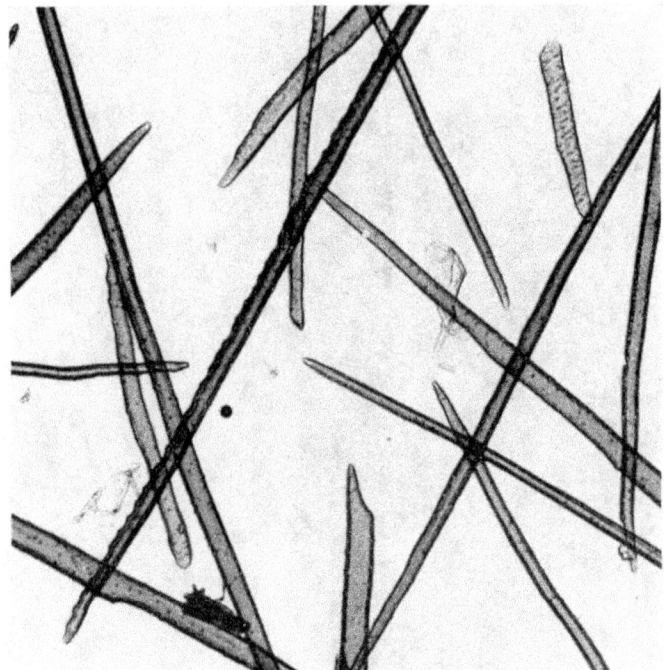

FIG. 8. Coir fibres (150x).

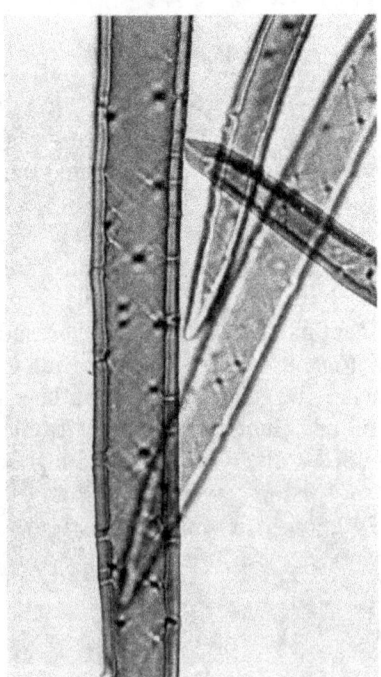

FIG. 9. Coir fibres: mid-portion, typical ends (500x).

BAST FIBRES

FLAX
(Linum usitatissimum)

FLAX is an annual that is cultivated in many parts of the world for both its fibre and for linseed oil. It grows to a height of about 3 feet and a diameter of 2 to 4 mm. The bast fibres of flax, known as linen fibres, are obtained in bundles from the bark of the stem. Separation of these fibres from other cellular elements of the short stem is achieved by "retting," a process of rotting, and then running the stems through breaker rolls. In this way the stems are broken without damaging the fibres. The unwanted broken matter is separated from the fibres by a beating and combing process. Flax was probably the first plant fibre to be used by man for textiles. Linen wrappings on the Egyptian mummies have been identified as being up to 4,500 years old.

The flax fibre is used mainly for strong fabrics, industrial sewing threads, twines, etc. It is also a constituent of the rag papers produced from textile waste and old rags. These papers are used for a variety of goods, including currency papers, map papers, carbon papers, blueprint papers, and felt papers. Flax stems are also used directly as a raw material in pulp mills. Airmail and cigarette papers are made from the flax stems after they have been broken down and pulped. In addition to the bast fibres these papers contain short elements from the bark and the woody stalk. These elements are commonly known as shives. They stain blue with Herzberg stain, while the bast fibres stain a reddish brown.[2]

A single flax fibre is of cylindrical shape with a smooth surface (Figs. 10 and 11). The diameter is almost uniform in the central portion, tapering gradually to a fine point at the ends. A very fine lumen, thick walls, and clear transverse dislocations are characteristic of flax fibres.[1] These dislocations, often called "nodes," are in the shape of I, X, or V, and are swellings (bulges) which are made more apparent by staining with methyl violet or Herzberg stain. The cross-section of a fibre is round to polygonal.

The nodes are often used in identification of flax fibres, although they are also

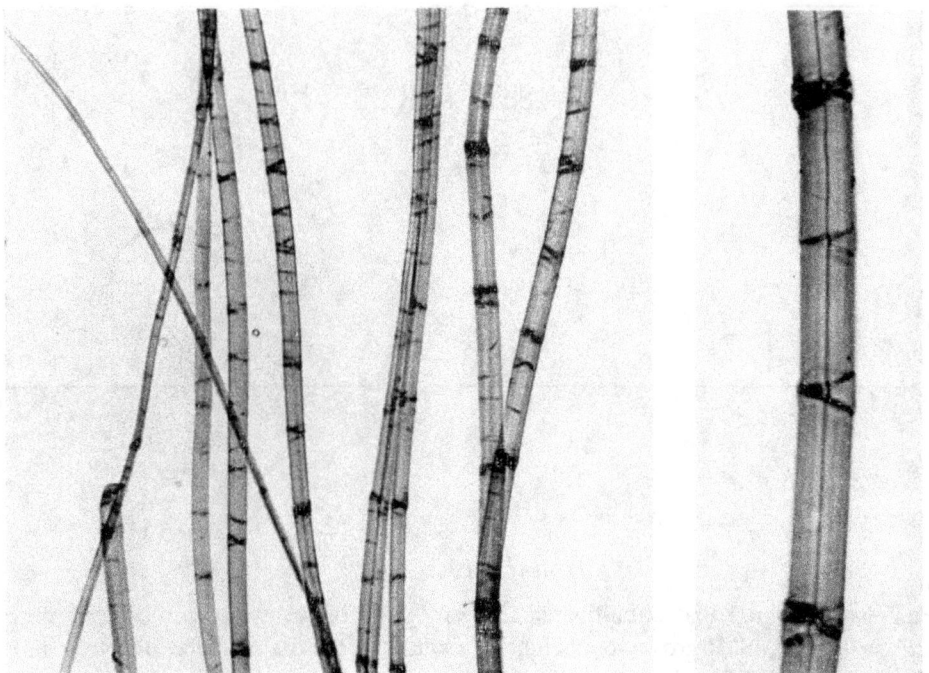

FIG. 10. Flax fibres (150x).   FIG. 11. Flax fibre (500x).

found in many other bast fibres. In these other fibres, however, the walls are thinner, and the nodes are not as conspicuous as in flax.[4]

Flax fibres readily defibrillate when beaten, and it is difficult to identify them in this state. However, both the fibrils and the fibres stain red with Herzberg stain, and this is a help in identification.[3] In addition, by using Herzberg stain with higher magnification, it can be observed that the lumen contains yellow-coloured protoplasmic particles. The narrow lumen is sometimes apparent only as a dark line, or it may even be completely closed at some places. This is a feature that distinguishes flax from the closely related hemp fibre whose lumen is always wide and continuous.[4] If a wet flax fibre is held at one end and allowed to dry, the free end (towards the observer) will be seen to move in a clockwise direction. Hemp fibre under similar conditions follows an anti-clockwise movement.[8]

The probable cause for a wide variation that occurs in flax fibre dimensions, as found by various investigators, is the large number of varieties of this species. An approximate range of dimensions is as follows:[8]

FIBRE LENGTH: 9 to 70 mm, average 33 mm.
FIBRE WIDTH: 5 to 38 microns, average 19 microns.

HEMP
(*Cannabis sativa*)

Hemp is an annual which is grown in the western hemisphere for textiles, cordage, oil seed, and paper. As there are several hemps this species is known as the "true hemp." It grows from 7 to 10 feet in height and about 12 mm in diameter. The fibre is obtained from the inner bark of the stem.

The fibre shape is cylindrical with longitudinal surface striations, transverse fractures, and swollen fissures (Figs. 12 and 13). These features are similar to flax but are not so pronounced in the case of hemp. The fibre walls are thick, while the lumen is broad, flat, and narrows toward the end of the cell. The fibre diameter is markedly uneven. The fibre ends are blunt, thick-walled, and occasionally have lateral branches. The cross-section of a fibre is polygonal with rounded edges.[1] Rare fragments of parenchymatous tissue, torn-shaped epidermal hairs, as well as epidermal cells themselves, may also be found with the fibres. They stain yellow with Herzberg stain.[3, 4]

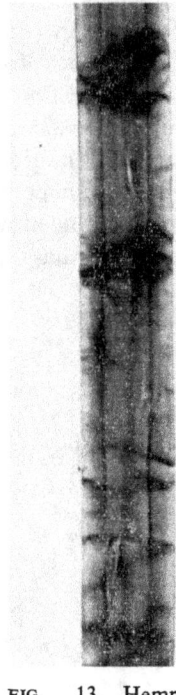

FIG. 12. Hemp fibres (150x).   FIG. 13. Hemp fibre (500x).

There is a great similarity between hemp and flax fibres, and it is not easy to differentiate between them. However, hemp fibres may be distinguished because they have wider cell lumen and more rounded fibre ends.[6] Another means of differentiation is to boil the fibres in commercial cyanine solution. As the middle lamella of hemp is more lignified, the fibres stain a green-blue colour. The less lignified flax fibres do not stain at all.[4]

With Herzberg stain, hemp fibres stain blue or violet with traces of yellow, according to the chemical treatment they have received.

FIBRE LENGTH:[8] 5 to 55 mm, average 25 mm.

FIBRE WIDTH: 10 to 51 microns, average 25 microns.

JUTE
(Corchorus capsularis)

Jute is one of the important bast fibres. The fibres are obtained in strands from the bark of herbaceous annuals of the genus *Corchorus*. The principal cultivated species are *C. capsularis* and *C. olitorius*, both of which have a large number of varieties.

Jute is grown in East Pakistan, India, and Brazil, amongst other countries. The height of the plant ranges from 5 to 16 feet and the diameter of the stalks is about 10 to 20 mm. Jute is chiefly used for the manufacture of cordage, bagging, and wrapping materials.[1] Jute enters the paper mills largely as new burlap cuttings and old sugar bagging.

A characteristic feature of the jute fibre is the irregular width of the broad and well defined lumen (Figs. 14 and 15). Sometimes the lumen closes up and is entirely missing for a short distance.[4] The individual fibre is cylindrical with little variation in diameter. The fibre walls are thick and generally smooth, having more or less numerous nodes and cross-markings depending upon the mechanical treatment they have received. The fibre ends are slender and pointed. The cross-section of a fibre is of polygonal shape with sharply defined angles and a round or oval lumen.[1] The fibres are surrounded with a layer of lignin material. Because of the heavy lignification, jute that has not been drastically cooked and bleached always stains

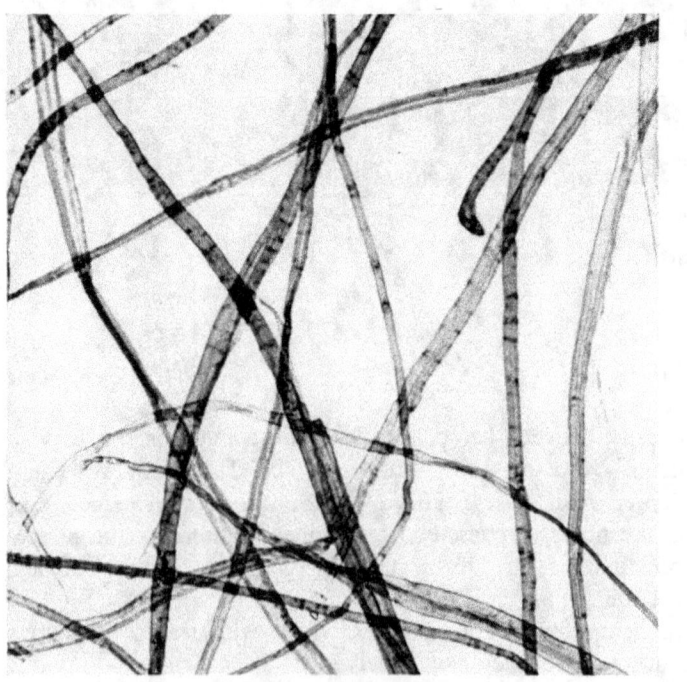

FIG. 14. Jute fibres (150x).

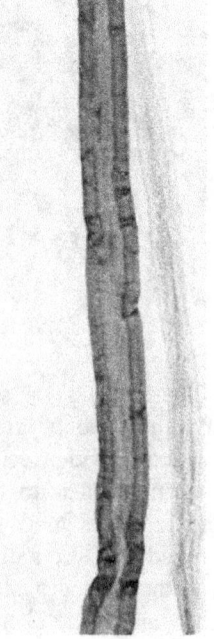

FIG. 15. Jute fibre (500x).

yellow-green or yellow with Herzberg stain. In various furnishes, jute fibres are often found in bundles, and this is of help in identification.

FIBRE LENGTH:[7]   1.5 to 5 mm, average 2 mm.

FIBRE WIDTH: 10 to 25 microns, average 20 microns.

RAMIE
(Boehmeria nivea)

Ramie is a perennial plant growing to a height of 5 to 8 feet and having a diameter of from 12 to 20 mm. It is cultivated mainly in the United States, Asia, and Africa. The fibres, which are remarkable for their size, are obtained from ramie bark and are mainly used for fish nets, canvas, industrial packings, and upholstery fabrics. The waste and short fibres are used for the manufacture of paper.

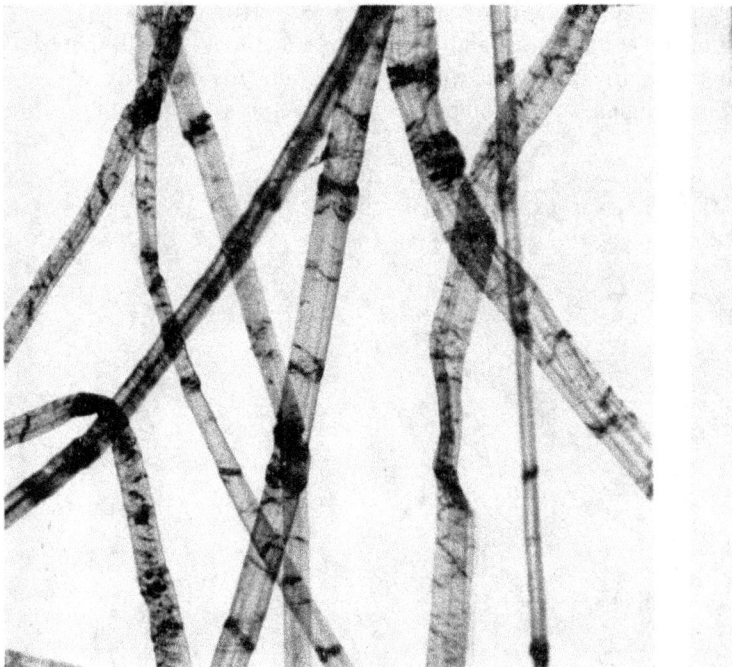

FIG. 16. Ramie Fibres (150x).

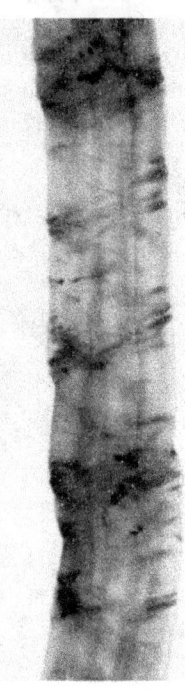

FIG. 17. Ramie fibre (500x).

The commercial ramie fibre consists of individual fibres similar to those of flax and hemp, but they are much larger in all dimensions, which alone are sufficient for identification purposes. The width of a single fibre often varies throughout its length, ranging from sections which are flat and wide to others which are narrow with thick walls (Figs. 16 and 17). The lumen is rather well defined but not very wide. The fibre tapers off to a narrow rounded end,[7] and in cross-section the fibre is elliptical.[5]

Transverse markings and striations are of frequent occurrence, and they are finer than those of flax. They stand out better in polarized light.[3] The lumen often contains some granular material and remnants of protoplasm. Ramie fibres are composed of nearly pure cellulose. They stain brownish-purple, blue, or red, with Herzberg stain.

FIBRE LENGTH:[6] 60 to 250 mm, average 120 mm.
FIBRE WIDTH: 11 to 80 microns, average 50 microns.

SUNN OR BENARES HEMP
(Crotalaria juncea)

Sunn is an annual which grows 8 to 10 feet tall with stalks 10 to 20 mm in diameter. It is cultivated in India, Pakistan, and the United States as a fibre plant and also for fodder and green manure. The fibres are obtained from the bark of the stem and are mainly used for twine, rug yarns, and papermaking.

Sunn fibres are cylindrical with surface striations and cross-markings (Figs. 18 and 19). The width of the lumen varies and most often contains yellowish material. The fibre ends are irregularly thickened, blunt, and rounded. There is a comparatively thick layer of lignin surrounding each fibre. A cross-section of a fibre is of oval shape.[1,10]

FIBRE LENGTH:[1] 4 to 12 mm, average 7.5 mm.
FIBRE WIDTH: 25 to 50 microns, average 30 microns.

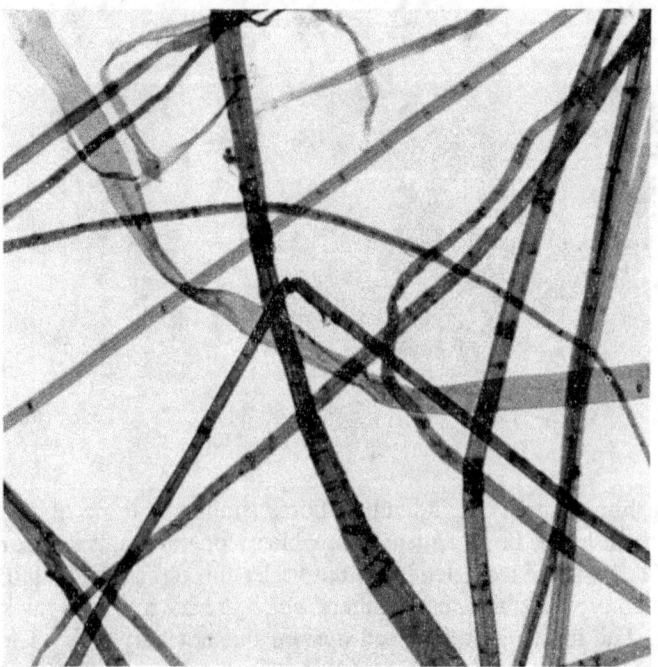

FIG. 18. Sunn fibres (150x).

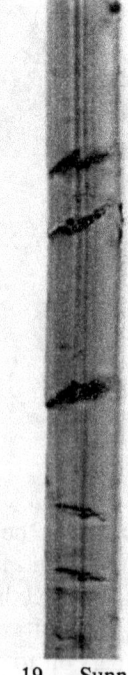

FIG. 19. Sunn fibre (500x).

KENAF
(*Hibiscus cannabinus*)

Kenaf (also known as Ambari or Gambo hemp) is an annual that grows to a height of 8 to 12 feet and to a diameter of about 12 mm. It is cultivated in India, Pakistan, Cuba, and the United States. The fibres, which are obtained from the bark of the stalk, are mainly used for cordage, canvas, and bagging.

Individual fibres are cylindrical with some surface irregularities and cross-markings (Figs. 20 and 21). The cell wall is thick and surrounded by a thick median layer of lignin. The lumen is irregular in width, has noticeable contractions, and at some points becomes discontinuous. Fibre ends are irregularly thickened and blunt. Fibre cross-sections are polygonal.[1,10]

FIBRE LENGTH:[1]   2 to 6 mm, average 5 mm.
FIBRE WIDTH: 14 to 33 microns, average 21 microns.

FIG. 20. Kenaf fibres (150x).

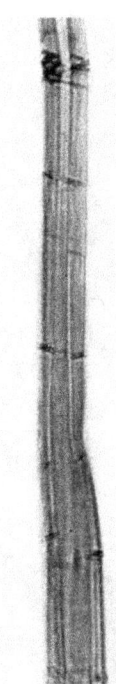

FIG. 21. Kenaf fibre (500x).

PAPER MULBERRY
(*Broussonetia papyrifera*)

This tree (also known as Kozo or Kodzu) grows in China, Japan, and the Pacific Islands. The fibres are obtained from the inner bark and are of two shapes (Figs. 22 and 23): (1) thick fibres with sharply pointed ends, and (2) ribbon-like fibres, which are often twisted and enveloped in a thin, transparent cuticle and have broad and round ends. The cuticle is typical of paper mulberry. It is visible in ordinary transmitted illumination, though it can be better observed if the fibres are examined in partially polarized light.

The lumen is small and difficult to distinguish because it is filled at intervals with yellowish material.

Small prismatic crystals of calcium oxalate are sometimes present in the fibres. Rare parenchymatous cells and some other bark elements may also accompany these fibres.[1,2,3]

FIBRE LENGTH:[2]   6 to 20 mm, average 10 mm.
FIBRE WIDTH: 25 to 35 microns, average 30 microns.

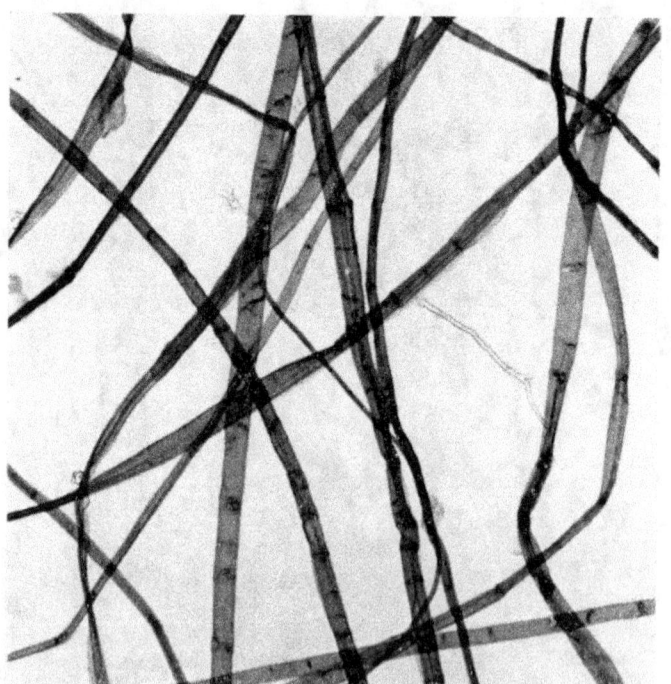

FIG. 22. Fibres of paper mulberry (150x).

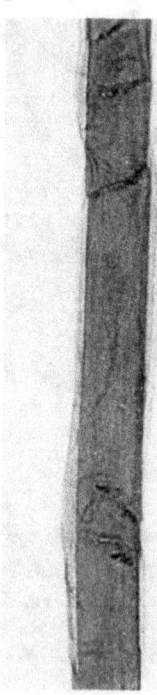

FIG. 23. Paper mulberry: ribbon-like with cuticle (500x).

MITSUMATA
(Edgeworthia papyrifera)

Mitsumata fibres are obtained from the bark of a shrub cultivated in Japan for its fibre. The fibres are not long, but they are slender and are used for papermaking.

A diagnostic feature which distinguishes mitsumata fibres from other known plant fibres is a fairly wide central segment about 0.3 mm long.[7] Beyond this segment the fibre is relatively narrow, appears slenderer, and tapers off to usually spatulate, but sometimes pointed or forked, ends.[3] The lumen is relatively wide in the central portion and quite narrow beyond it (Figs. 24, 25, 26). The fibre walls are occasionally marked with relatively indistinct cross-markings. Most frequently these bast fibres are not entirely separated from other cells of the phloem and therefore are accompanied by rare rectangular parenchyma cells and sieve-tube elements.[7]

FIBRE LENGTH:[7]   2.4 to 3.6 mm, average 2.9 mm.

FIBRE WIDTH OF CENTRAL SEGMENT: 11.9 to 27.3 microns, average 18.4 microns.

FIBRE WIDTH BEYOND CENTRAL PORTION: 6.8 to 13.9 microns, average 9.0 microns.

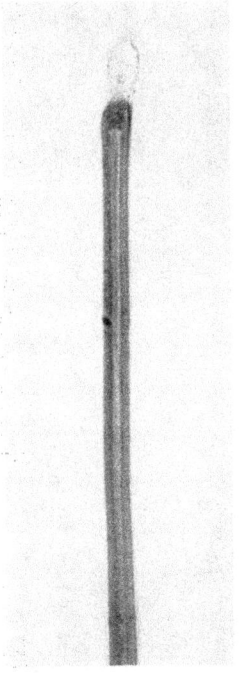

FIG. 24. Spatulate end of a mitsumata fibre (500x).

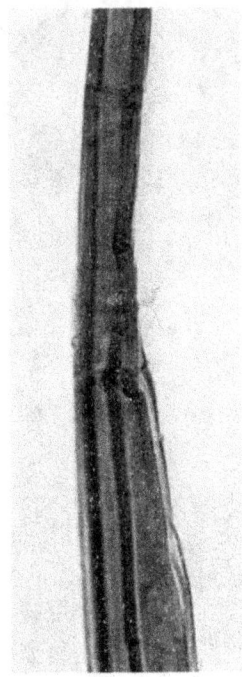

FIG. 25. Mitsumata fibre (500x).

FIG. 26. Mitsumata fibres (150x).

LEAF FIBRES

MANILA HEMP OR ABACA
(Musa textilis)

MANILA HEMP or abaca is a perennial that is a variety of the banana plant indigenous to the Philippines and Indonesia. In height it ranges from 15 to 35 feet and it has a diameter of 50 mm or less. Its stem is surrounded by leaf sheaths arising at the base of the plant and enfolding the stem. The fibres are obtained in the form of strands from the vascular bundles of these sheaths. The main use of abaca fibres is for cordage. Waste and old rope are used for the production of strong and thin paper.[1]

Individual fibres are of fairly uniform diameter, are rather thin-walled, and usually have pointed ends (Figs. 27 and 28). The lumen is well defined, broad, of uniform width, and continuous. Many fine cross-markings are evident on most of the fibres. The longitudinal shape is approximately cylindrical. Its cross-section varies from an irregular oval to a polygon with rounded corners. In cross-section the lumen is rounded, large, and well defined. Cell walls are thin in relation to the area of cross-section.[1]

Manila fibre bundles often show thick, strongly silicified plates (stegmata). They may be observed after a strand has been macerated in a chromic acid solution, extracted with nitric acid, and ignited. If the ash is examined under the microscope after treatment with dilute acid, the stegmata appear like a string of pearls. These often occur in long chains.[1] Some accompanying cells and vessel segments may also be found in manila furnishes.[6]

FIBRE LENGTH:[8]   2 to 12 mm, average 6 mm.
FIBRE WIDTH: 16 to 32 microns, average 24 microns.

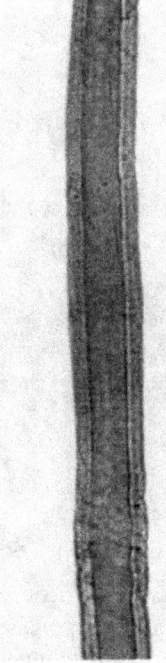

FIG. 27. Manila fibres (150x).   FIG. 28. Manila fibre (500x).

SISAL
(*Agave sisalana*)

Sisal fibres are obtained from the leaves of Agave sisalana, a perennial of which there are several varieties. Sisal is cultivated in Africa, Central and South America, and also in the West Indies. The plant has a short stem with leaves from 4 to 6 feet long. The commercial fibre, which is in the form of strands, is used for ropes and twines and other applications. Sisal fibres found in the paper industry probably originate from waste material.

Individual fibres of sisal are nearly cylindrical in shape (Figs. 29, 30, 31). Cell walls are thick and have some fine cross-markings. The lumen is usually large, not too prominent, and varies in width. The fibre ends are generally thick and blunt, but are sometimes pointed or forked. The cross-section of a fibre is irregularly oval to rounded polygonal.[1] Occasionally sisal fibres are accompanied by spiral vessels, short and thick-walled fibres (with distinct surface pores), and parenchymatous cells. The latter cells contain single calcium oxalate crystals, which resemble a thick needle in form and are often quite long. After combustion they glisten in the ash.[1]

The fibre dimensions cited by various authors differ considerably, though on the average they seem to be as follows:[8]

FIBRE LENGTH: 0.8 to 8 mm, average 3.3 mm.
FIBRE WIDTH: 8 to 41 mircons, average 20 microns.

LEAF FIBRES

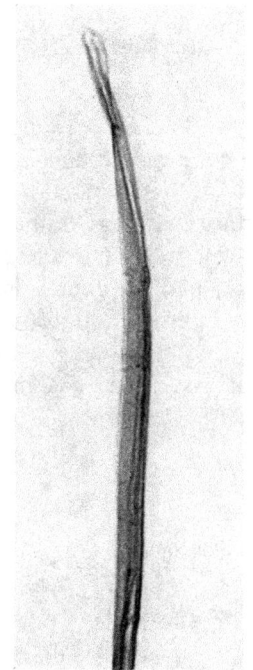

FIG. 29. Typical end of a sisal fibre (500x).

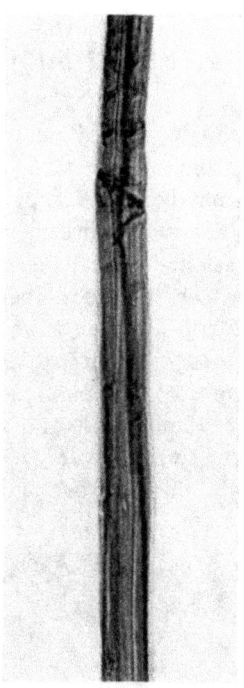

FIG. 30. Sisal fibre (500x).

FIG. 31. Sisal fibres (150x).

NEW ZEALAND FLAX
(Phormium tenax)

This plant is indigenous to New Zealand. It has leaves from 3 to 8 feet long. The fibres are obtained from the leaves and are mainly used for cordage and bagging fabrics. However, they may be found in paper.

Individual fibres are nearly cylindrical, without evident surface markings, rather smooth, and have occasional wave-like irregularities in the cell walls (Figs. 32 and 33). The fibre ends are pointed. The cell walls are thick, and the lumen is uniformly narrow and often distinguishable only with difficulty. In cross-section the fibre shape is approximately circular.[1]

The fibres, when incompletely separated from the other leaf tissues, are accompanied by vessel segments, epidermal, and parenchymatous cells.

FIBRE LENGTH:[8]   2 to 15 mm, average 7 mm.
FIBRE WIDTH: 5 to 27 microns, average 15 microns.

FIG. 32. New Zealand flax (150x).

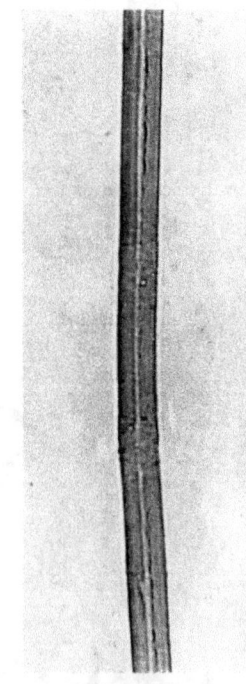

FIG. 33. New Zealand flax fibre (500x).

PINEAPPLE
(Ananas comosus)

Pineapple fibres are obtained from the leaves of the pineapple plant, which is cultivated for its well-known fruit in the Hawaiian and West Indian Islands. The leaves grow up to 4 feet long. The fibres are mainly for twine, thread, and fabrics.

LEAF FIBRES

Pineapple fibres are readily distinguishable from other fibres because they are extremely fine (Figs. 34 and 35). The longitudinal shape of individual fibres is approximately cylindrical with pointed ends. The fibre surface is smooth. The lumen is very narrow and appears almost like a line. The cross-section of a fibre is oval to rounded polygonal and frequently it is flattened. Chemically, pineapple fibres have a higher percentage (81.5) of cellulose than most of the leaf fibres.[1]

FIBRE LENGTH:[8] 1.7 to 10 mm, average 5 mm.
FIBRE WIDTH: 3 to 13 microns, average 6 microns.

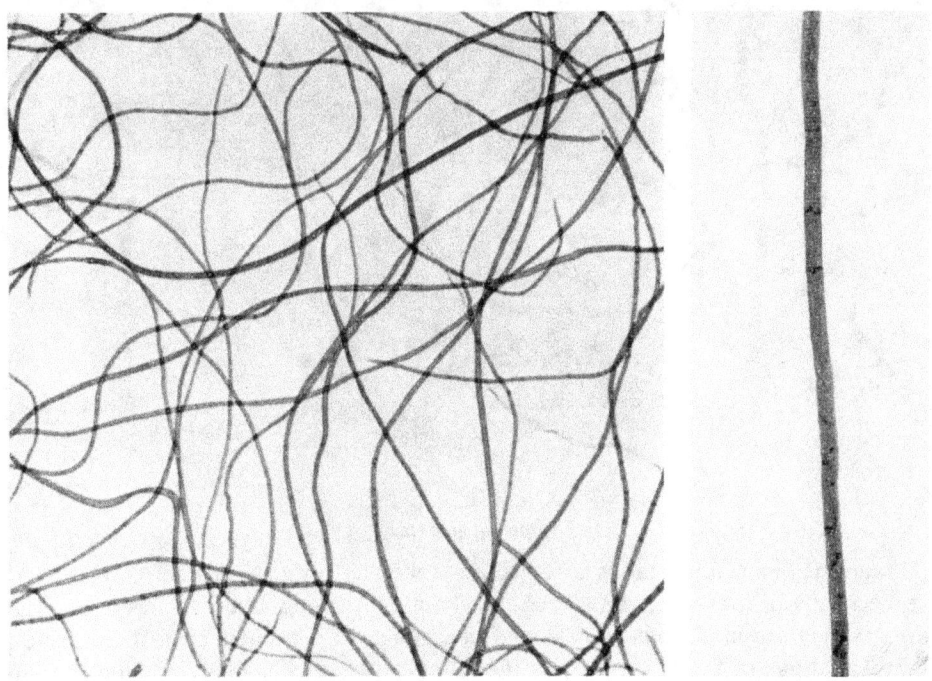

FIG. 34. Pineapple fibres (150x).   FIG. 35. Pineapple fibre (500x).

CAROA
(Neoglazovia variegata)

Caroa fibres are obtained from the leaves of the wild plant, caroa, which grows abundantly in Brazil. The leaves grow up to 12 feet long. The fibres are used in the manufacture of cord, rope, twine, rayon, cloth, and paper.

Caroa fibres (Figs. 36 and 37) resemble those of pineapple and pita floja (not described here). They particularly fit the description of pineapple fibres. Chemically, caroa fibres contain a low percentage (58.9) of cellulose and a relatively high percentage (12.7) of lignin.[1,2]

FIBRE LENGTH:[2] 2 to 6 mm, average 4 mm.
FIBRE WIDTH: 8 to 12 microns.

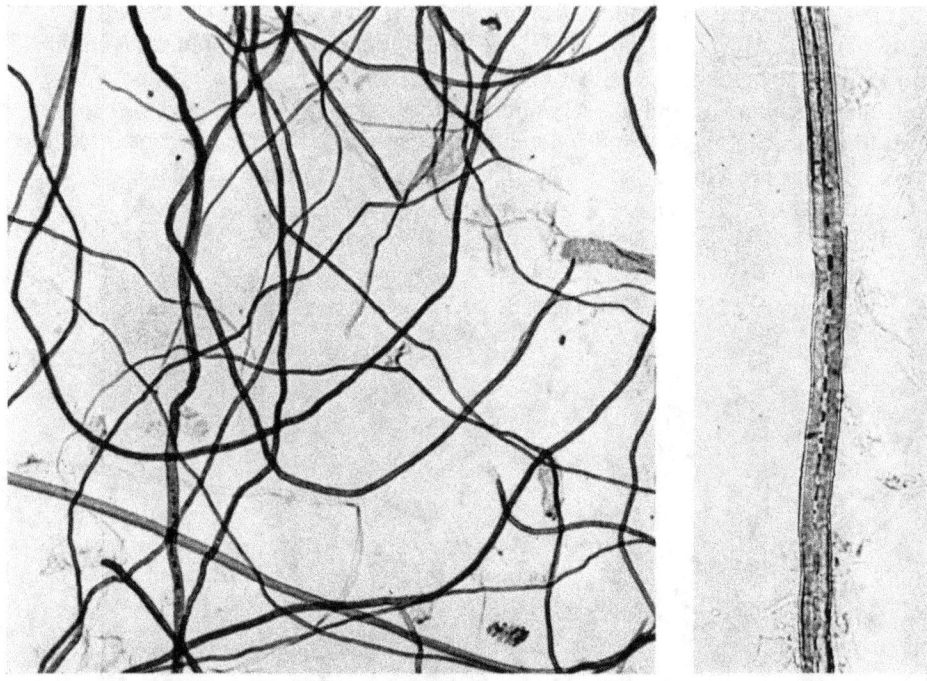

FIG. 36. Caroa fibres (150x).

FIG. 37. Caroa fibre (500x).

MAURITIUS HEMP
(Furcraea gigantea)

Mauritius hemp, indigenous to Brazil, is cultivated in East Africa, Ceylon, and on the Island of Mauritius. Its leaves are about 8 feet long and 8 inches wide. The fibres, which are in strands from 4 to 7 feet long, are obtained from these leaves.

Principal use of the fibres are for the manufacture of bagging and other coarse textiles. Like many other textile fibres they also find their way into paper furnishes.

The longitudinal shape of individual fibres is approximately cylindrical (Figs. 38 and 39); the cross-sectional shape is polygonal. The lumen is very large, while the fibre wall is relatively thin, smooth and lacks any special markings or characteristics. The fibre ends are blunt.[1]

There are nine other species of the genus *Furcraea* from which fibres are obtained. They grow wild or are cultivated in the Caribbean and Central and South America.

FIBRE LENGTH:[8]   1.3 to 6 mm, average 2.9 mm.

FIBRE WIDTH: 18 to 32 microns, average 23 microns.

FIG. 38. Mauritius hemp fibres (150x).

FIG. 39. Mauritius hemp fibre (500x).

THE PULPS OF GRASSES

STRAWS

THE GRASSES described in this section belong to the family Gramineae. The entire plant, stem, and the leaves are used as raw materials for pulping, and as a result grass pulps contain fibres as well as a variety of other cell types. As all grasses are closely related their cellular elements are very similar. Variations in the size of cells is often the main distinguishing feature of the genera. Wheat straw, which is the most commonly found grass pulp, has been chosen to illustrate the various cellular elements found in grass pulps (Figs. 40 to 48).

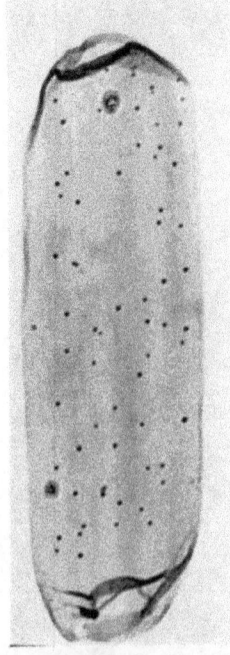
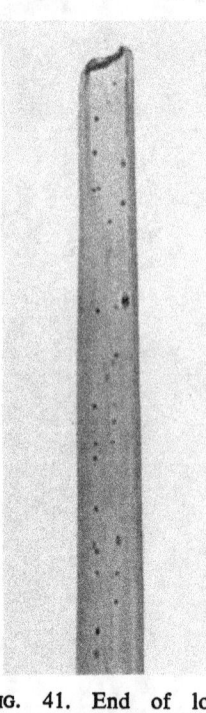
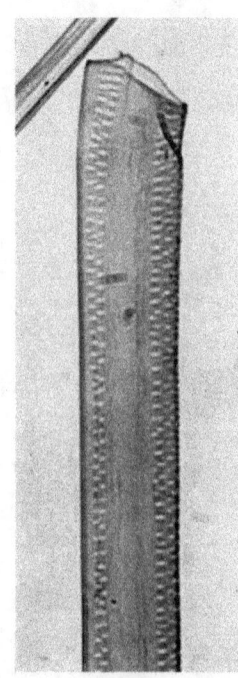

FIG. 40. Broad (barrel-shaped) parenchyma cell (300x).

FIG. 41. End of long parenchyma cell (300x).

FIG. 42. End of long pitted vessel segment (300x).

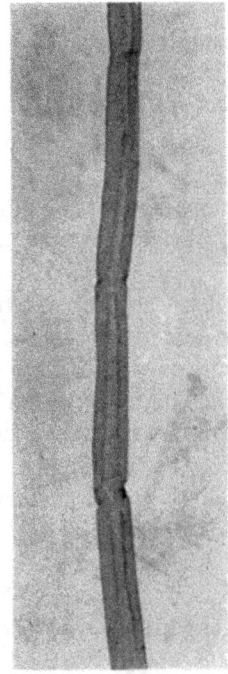

FIG. 43. Mid-portion of a fibre (300x).

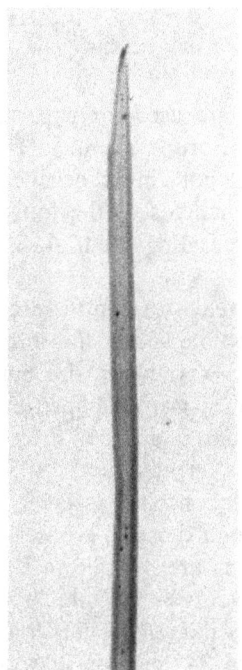

FIG. 44. End of a fibre (300x).

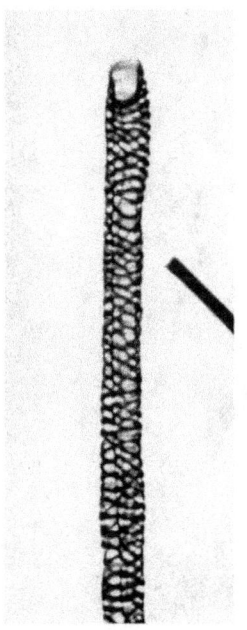

FIG. 45. End of a tracheid (300x).

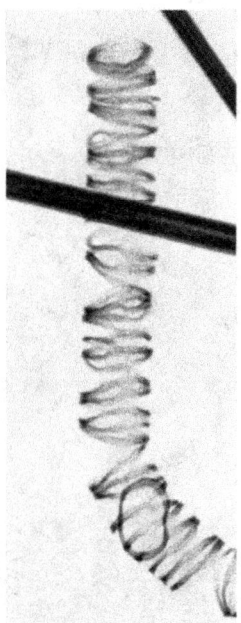

FIG. 46. Portion of a vessel segment with annular thickenings (300x).

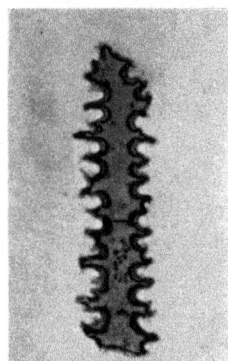
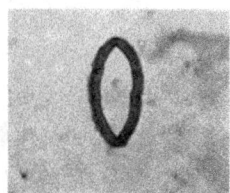

FIG. 48. Ring-thickening from an annular vessel segment (300x).

CEREAL STRAWS

Wheat (*Triticum sativum*) Barley (*Hordeum* spp.)
Oat   (*Avena sativa*)   Rye   (*Secale cereale*)

Cereal straws are used for papermaking in areas where wood is not abundant. For example in Europe chemical pulps of straws are used in the production of various grades of both bleached and unbleached paper or board.

The cellular elements of the four cereal straws are not identical, but there are difficulties in separating them. It is more practical to review them as a group (Figs. 49 to 52).

In general, cereal straw fibres are longer and wider than those of esparto. The fibre ends are pointed, and the lumens may vary from broad to narrow. Vessel segments are very variable in size but never reach the maximum diameter of either sugar cane or corn. Barrel-shaped and cylindrical parenchyma cells of the ground tissue are abundant. Epidermal cells vary in size and form, are sparsely pitted, and have more or less serrated margins. The guard cells which surround the stomata are smooth-walled. Inconspicuous erect trichomes are also present, though they occur less frequently than those in esparto and rice.[6,7]

Cell dimensions are as follows:[7] typical thick-walled fibres range from 0.68 to 3.12 mm and average 1.48 mm in length; the width is from 6.8 to 23.8 microns and average 13.3 mircons. The very thin-walled fibres vary from 0.8 to 2.9 mm in length and 27 to 34 microns in width. Parenchyma cells are up to 0.45 mm in length and 130 microns in width. Vessel segments: the maximum dimensions are 1.0 mm in length and 60 microns in width. Epidermal cells are from 0.036 to 0.45 mm in length and from 10 to 41 microns in width. The latter are of very variable form, rhombic or rectangular in outline.

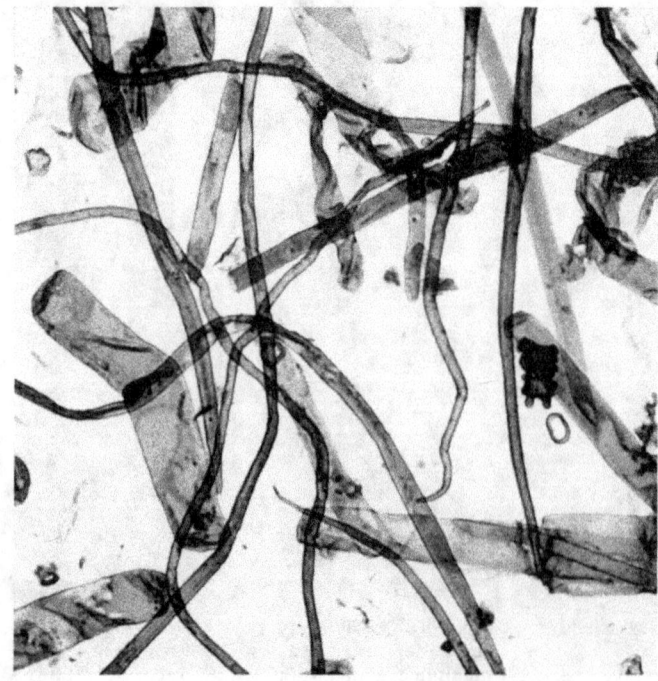

FIG. 49. Wheat straw pulp (150x).

FIG. 50. Barley straw pulp (150x).

FIG. 51. Oat straw pulp (150x).

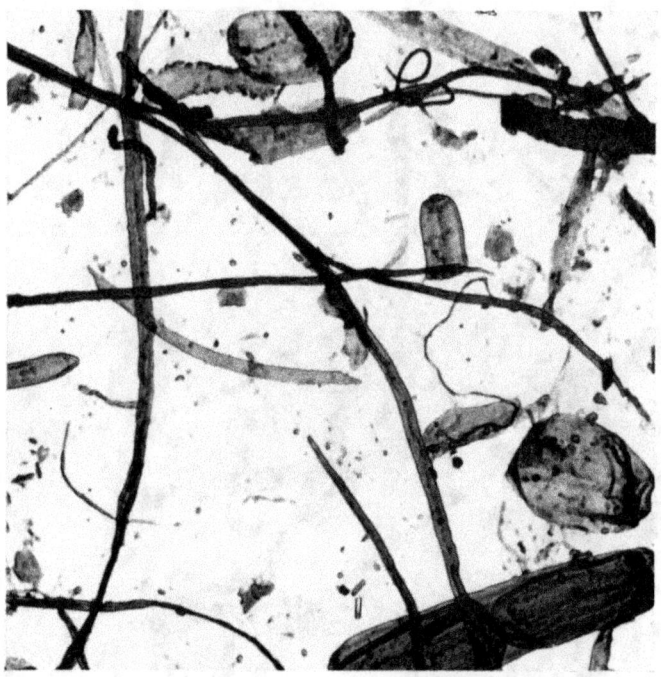

FIG. 52. Rye straw pulp (150x).

RICE
(*Oryza sativa*)

Rice straw elements closely resemble those of esparto grass but are distinguishable from the elements of cereal straws, sugar cane, corn, and bamboo, largely because they are much smaller (Fig. 53).

The fibres of rice are longer and appear slenderer than those of esparto. The epidermal cells are serrated, heavily pitted, and are usually present in clusters. Among these cells are trichomes of about the same size as those of esparto. Rice trichomes may be distinguished from those of esparto by their straight tips. The stomata are surrounded by conspicuously dentate guard cells.[6,7]

The fibre dimensions are as follows: length, 0.65 to 3.48 mm, average 1.45 mm; diameter 5.1 to 13.6 microns, average 8.5 microns. The barrel-shaped parenchyma cells (typical for all the grass species except esparto) and the cylindrical ones of the same type are up to 0.35 mm long and 82 microns wide. Vessel segments are slender and up to 0.65 mm long and 40 microns wide. The trichomes, which have short tips, are 38 to 45 microns long and 17 microns wide.

SUGAR CANE BAGASSE
(*Saccharum officinarum*)

Bagasse is the refuse left after the crushing of sugar cane. It is being used more and more as a raw material for making paper and hardboard products.

The fibres from the vascular bundles are thin to thick walled (Fig. 54) with pointed ends and no special surface markings except for the presence of occasional small pits. Thin-walled, fibre-like short cells, with oblique, blunt, or even forked ends may also be found. The most remarkable feature of sugar cane bagasse pulp is the abundance of large parenchyma cells and vessel segments. Parenchyma cells are thin walled and pitted. Vessel segments are of all types as in the other grasses, although they are distinctive because of the great length of the largest ones. Epidermal cells are infrequent; they are narrow, rectangular, and have slightly serrated margins. The guard cells of the stomata are smooth-walled.[6,7]

Fibre dimensions:[7] length, 0.8 to 2.8 mm, averaging 1.7 mm; width, 10.2 to 34.1 microns, averaging 20 microns. Parenchyma cells: length, up to 0.85 mm; width, up to 140 microns. Vessel segments: length, up to 1.35 mm; width, up to 150 microns.

CORN
(Zea mays)

Cornstalk pulp has been used for the production of paper and building board materials but only on a small scale.

The cells of cornstalks (Fig. 55) are not readily distinguishable from those of sugar cane. There is a great similarity between them, and the description of sugar cane elements also applies to cornstalks. The main difference between the various cells of the two species is their size. The parenchyma cells and vessel segments of cornstalks are shorter than those of sugar cane.

Typical fibres are from 0.5 to 2.9 mm (average 1.5 mm) in length and 14 to 24 microns (average 18 microns) in width. The shorter thin-walled fibres average 0.44 mm in length and 16.4 microns in width. Parenchymatous cells are up to 0.33 mm long and 150 microns wide. Vessel segments are up to 0.6 mm long and 150 microns wide. Epidermal cells are sparse.[7]

ESPARTO
(Stipa tenacissima)

Esparto grass grows wild in North Africa and Spain where it is available in plentiful supply. Because of the fibre quality, esparto pulp is used in England for the manufacture of fine-quality book and medium-class writing papers.

Esparto fibres are slender, cylindrical, and taper off to pointed ends. They are thick or thin-walled, and the lumen is narrow and seldom detectable (Fig. 56). Esparto fibres are finer than those of wheat, rye, barley, and oats but not as fine as those of rice. Parenchyma cells and vessel segments are also much smaller than in the above mentioned species, excluding rice. Serrated epidermal cells are often present, though they are again smaller than in the cereal straws mentioned. The guard cells surrounding the stomata are smooth-walled.

The main feature which distinguishes esparto fibres from other grasses is the occurrence of small comma-shaped cells. These cells are known as trichomes and

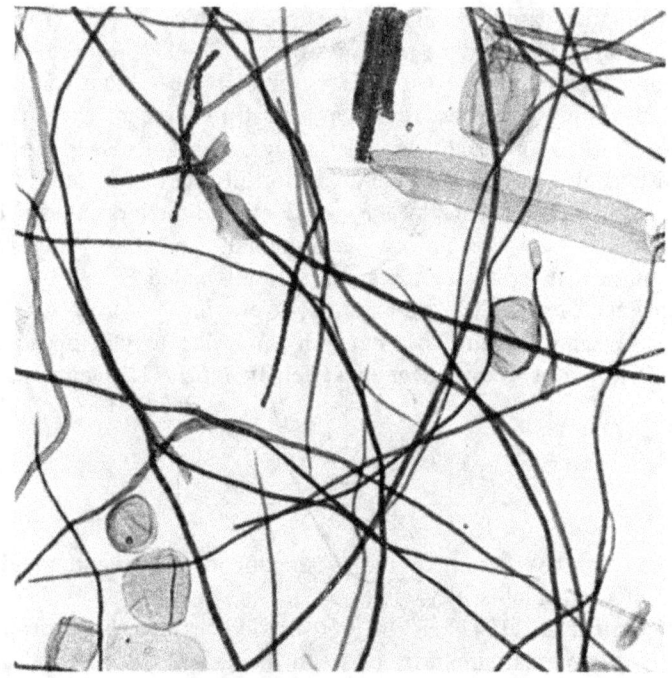

FIG. 53. Rice straw pulp (150x).

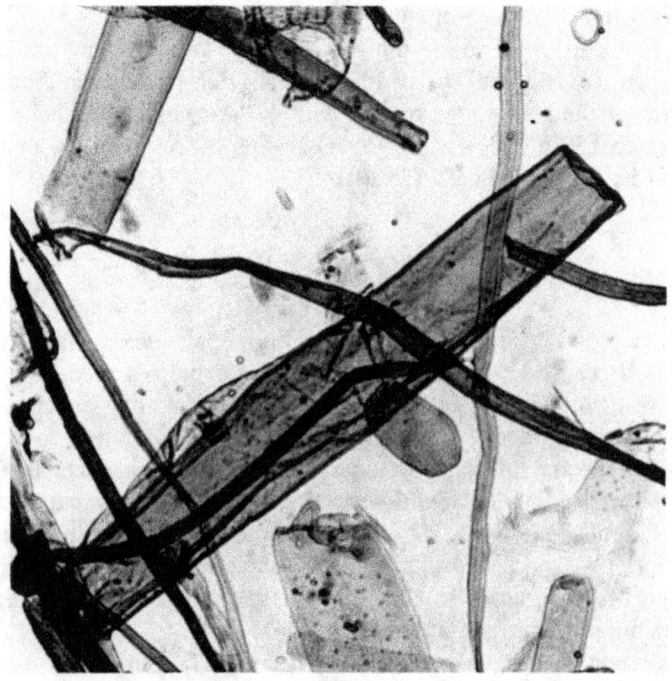

FIG. 54. Sugar cane bagasse (150x).

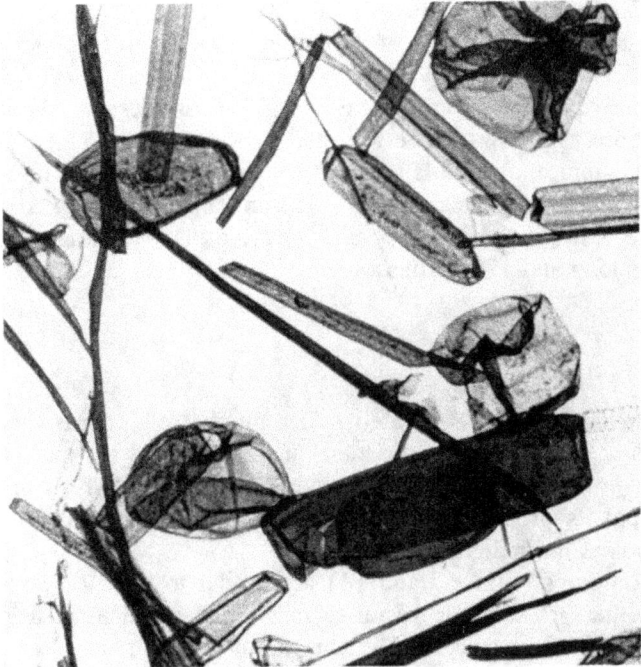

FIG. 55. Pulp of cornstalks (150x).

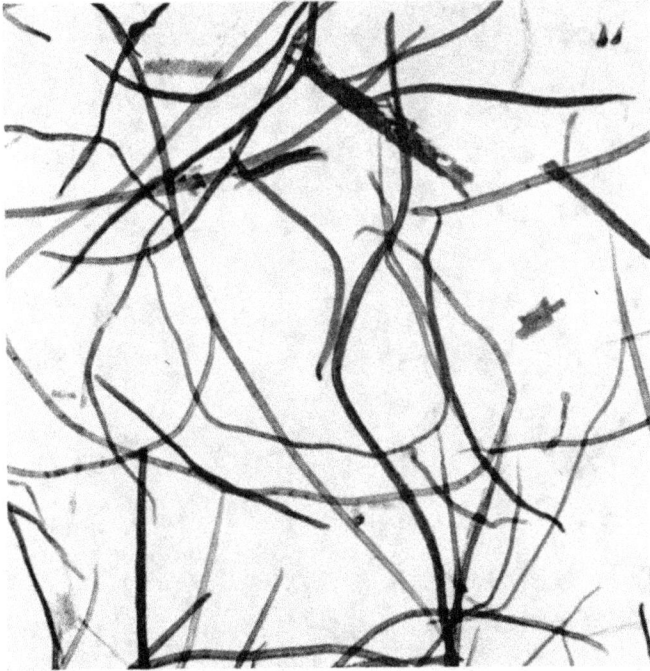

FIG. 56. Esparto fibres. Note the two comma-shaped cells (trichomes) in the upper right corner (150x).

are derived from the stiff hairs of the esparto leaves. Their form is rather similar to the comma mark in punctuation. Because of these trichomes esparto fibres are often known as "comma hairs."

Esparto fibre dimensions, quoted by various authors differ considerably. Those cited by F. F. Wangaard are as follows:[7] The fibre length is between 0.51 to 1.6 mm, with an average of 1.1 mm; the width is between 6.8 to 13.6 microns, with an average of 9.2 microns. Parenchyma cells are up to 0.35 mm long and 20.5 microns wide. Epidermal cells are between 6.8 to 13.6 microns wide. The trichomes are between 30 to 50 microns long and 17 microns wide.

BAMBOO
(*Bambusa* spp. *et al.*)

Bamboos are tropical plants that are separated into a number of genera which account for several hundred species. They are woody in the sense of hardness of their tissues, though structurally they are typical of the grass family to which they belong. Bamboo paper and its products are produced in the countries of South Asia where these giant grasses are indigenous.

The elements of bamboo pulps (Figs. 57 and 58) are similar to those of cereal straws, sugar cane, and other grasses. The vessel segments and parenchyma cells are smaller than those of sugar cane or cornstalks, but the fibres are much longer. Epidermal cells are not usually present.

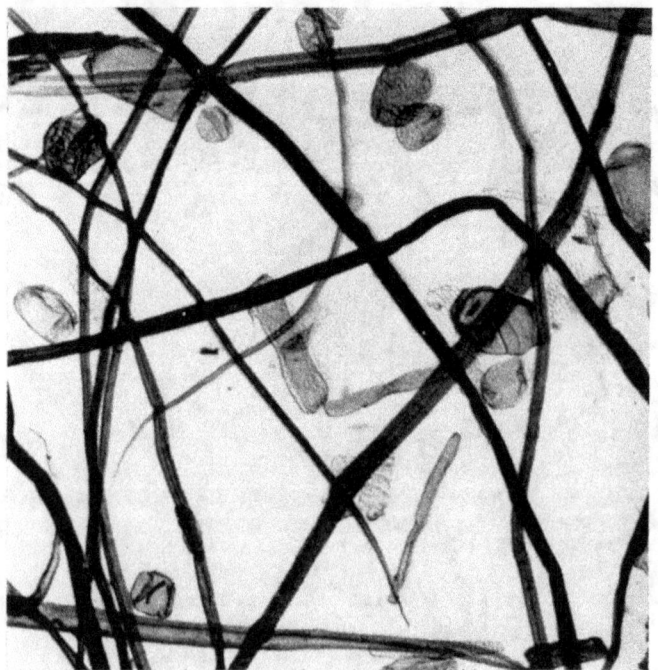
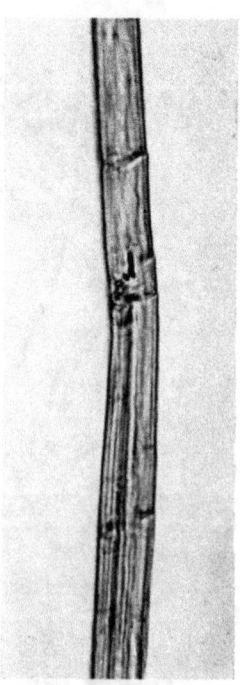

FIG. 57. Bamboo pulp (150x).

FIG. 58. Bamboo fibre (500x).

The dimensions of pulp elements of one of the bamboos (*Dendrocalamus arundinacea*) are as follows:[7] typical fibres are from 1.45 to 4.35 mm (average 2.7 mm) in length, and 6.8 to 27.3 microns (average 14.0 microns) in width. Then there are the wide thin-walled and pitted fibres which are 2.8 to 3.2 mm. long and 20.5 to 40 microns wide. The parenchyma cells are up to 0.25 mm long with a maximum width of 65 microns. The vessel segments are up to 100 microns wide.

ANIMAL FIBRES

WOOL AND HAIRS

WOOL FIBRES and the hairs of various other animals have basically the same structure and can be treated as a group for purposes of identification. They differ in appearance from all other paper fibres and are readily distinguishable under the microscope. However, their identification by animal origin is difficult since it depends on slight variations of general features which often overlap.

Wool is composed of a sulphur-containing protein, keratin. A growing wool fibre consists of a root and a shaft. The shaft usually consists of three layers: the epidermis, cortex, and medulla[11] (Figs. 59, 60, 61).

The epidermis, or the surface of the fibre, is made up of flat irregular horny scales which give the wool fibre its main diagnostic feature. These scales are arranged either shingle-like, overlapping longitudinally and circumferentially, or in a manner whereby the surface of the fibre is given a tile-like appearance. These different types of epidermis may be found on the same wool fibre.

The cortex is found below the protective epidermal scales. It constitutes the main body of the wool fibre and is made up of long, slightly flattened, and more or less twisted spindle-shaped cells.

The medulla or cellular marrow is found within the cortical layer and only in medium and coarse quality wool—the medullated fibres. It has a honeycomb-like structure. There is no medulla in fine quality wool fibres.

Wool is resistant to acids but deteriorates readily on treatment with alkalis, while cellulose fibres are resistant to alkalis but easily attacked by acids. Thus, the chemical properties of wool are opposite to those of cellulose.[11]

A 5 per cent solution of caustic soda at boiling temperature completely dissolves wool in a few minutes. Wool fibres are insoluble in cold acids. With nitric acid, wool is stained to a yellow colour which turns to an orange shade when it is subsequently treated with caustic alkali.[5]

The diameters of wool fibres vary greatly from 10 to 70 microns. There is not only a variation between fibres, but also a variation in diameter as well as shape

ANIMAL FIBRES 83

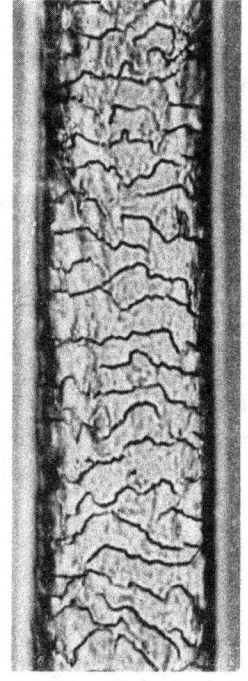

FIG. 59. Human hair (500x).

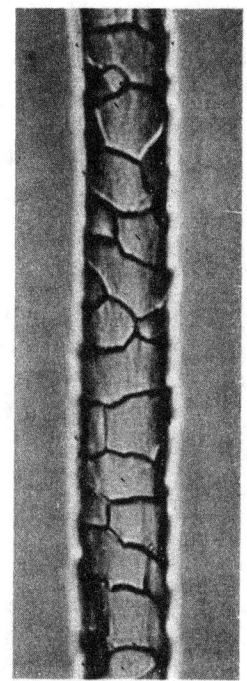

FIG. 60. Wool (500x).

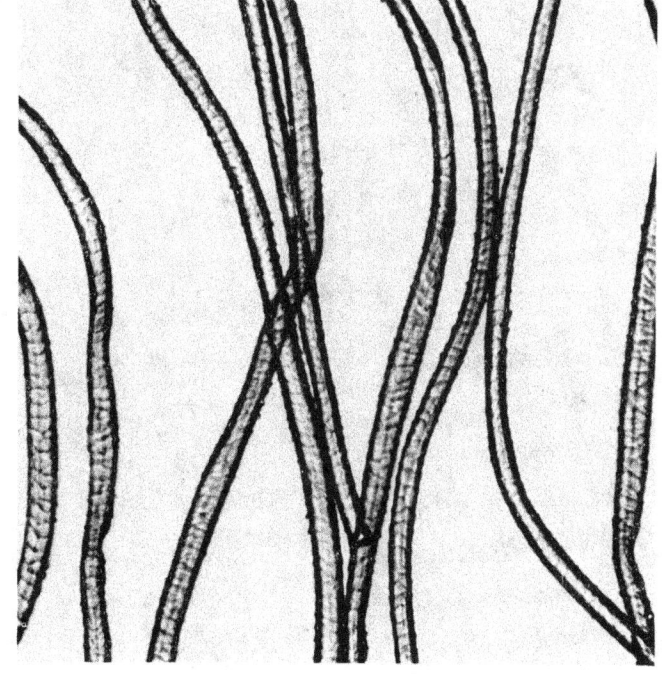

FIG. 61. Wool (150x).

along the length of individual fibres. Fine fibres are nearly circular in cross-section, while others are irregular in shape.

One type of wool fibre known as "kemp" fibre is normally short and wavy and tapers towards each end. It is of dead-white or opaque colour and is very coarse and brittle. The diameter of kemp fibres vary from 70 to 200 microns. They are found most frequently in carpet wool.[11]

SILK

The common silk is the filament from the cocoon spun by the larvae of the cultivated silkworm, *Bombyx mori*. It is not a true worm but a caterpillar which feeds on mulberry leaves.

On spinning the cocoon, the caterpillar secretes a viscous fluid and extrudes from two glands a pair of silk filaments which come together at a common exit in the caterpillar's head. The filaments emerge cemented together by a gum produced by another pair of glands. The filaments consist of the protein fibroin, which is elastic and strong. The cementing gum is of another protein, known as sericin, which is brittle when in dry state.[4]

The raw silk filament, as seen under the microscope, has an appearance which readily distinguishes it from other textile and paper fibres. The two component filaments, because of the sericin layer between them, often appear to be separated

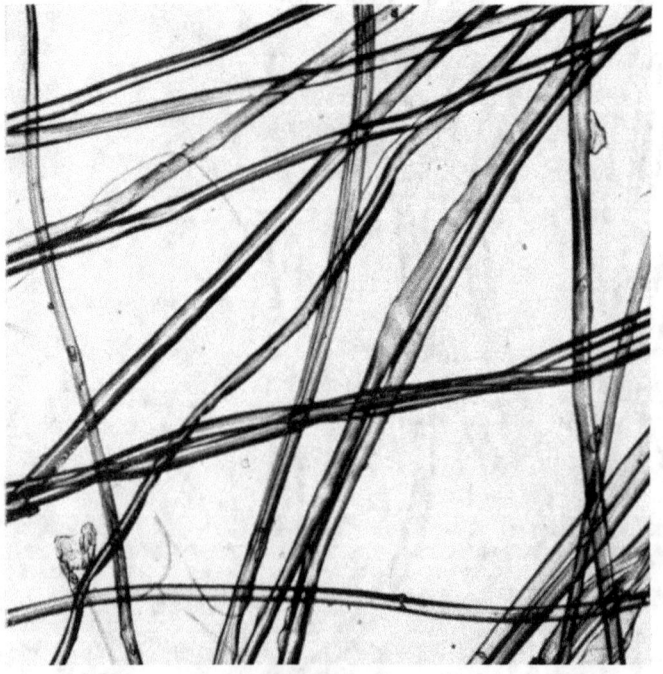
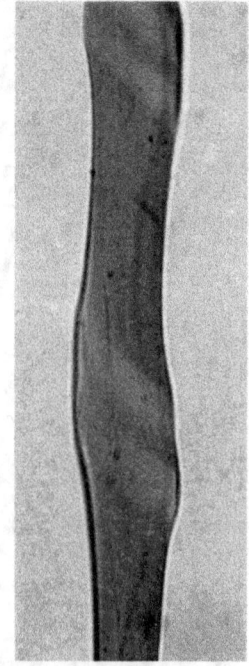

FIG. 62. Silk, degummed (150x).

FIG. 63. Silk, degummed (500x).

from one another by a gap. The surface of the raw silk fibre appears very irregular, with cross-fissures, creases, folds, as well as uneven lumps. These markings occur only in the sericin layer. They are not structural but are caused by the breaking and rubbing off of the gluey sericin layer during the reeling operation. Also, silk filaments appear as bundles of even numbers resulting from the reeling of the double filaments.[11]

The gum or sericin, which is from 18 to 23 per cent of the total weight, is removed by dilute alkali or hot soap solution. The degummed fibre shows a smooth, structureless, translucent filament with occasional constrictions as well as swellings or lumps (Figs. 62 and 63). Rare filaments are striated longitudinally, and the striations always run parallel to the fibre axis. In cross-section silk filaments are elliptical or triangular with rounded corners.[1] The diameter of a single fibre is usually from 10 to 12 microns but sometimes is up to 25 microns.[4]

Silk swells and dissolves in cold hydrochloric acid and in 80 per cent sulphuric acid. It turns yellow, swells, and disintegrates in concentrated nitric acid. Silk also gelatinises quickly and dissolves in boiling 5 per cent caustic soda, but not as readily as wool.[5]

TUSSAH SILK

There are several types of silk produced by wild silkworms of which Tussah silk (wild silk) is commercially the most important. The Tussah silkworm (*Antheraea pernyi*) feeds on leaves of a particular oak which grows in China and Manchuria.

Tussah silk is of tan colour and is noticeably coarser and stiffer than common silk (Figs. 64 and 65). The filaments are broad, ribbon-like, and show pronounced longitudinal striations—evidence that structurally the fibre is composed of fine fibrils. These fibrils may be isolated by maceration in cold chromic acid. The filaments also exhibit peculiar flattened cross-markings, caused by the overlapping of one fibre on another which occurs before the fibre substance has dried properly. Most frequently these markings run obliquely across the filaments and can be better observed under polarized light between crossed nicols. In cross-section Tussah filaments are typically wedge-shaped, exhibit a grainy inner structure, and a sawtooth-like contour of some of the fibres.[11]

Tussah silk is more resistant than common silk to acids and strong alkalis. It is almost unaffected by concentrated hydrochloric acid, chromic acid, and zinc chloride, all of which dissolve cultivated silk.[1] In boiling 5 per cent caustic soda it disintegrates to a pulp but does not dissolve completely.[5]

All wild silks are quite similar in their microscopic appearance, and it is rather difficult to differentiate between the various species. Wild silks collectively are distinguishable from cultivated silk in that they are of darker colour, ribbon-like form, striated longitudinal appearance, and are wedge-shaped in cross-section. The width of Tussah silk filaments varies from 9 to 51 microns, with an average of 28.5 microns.[11]

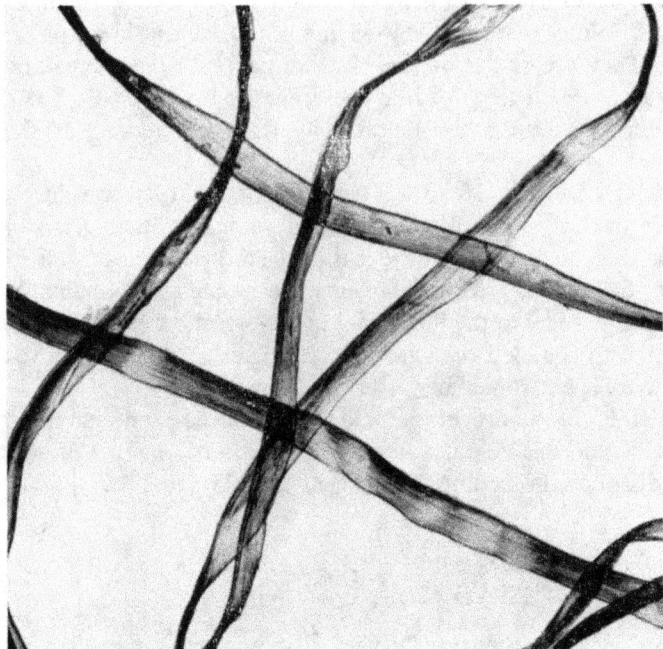
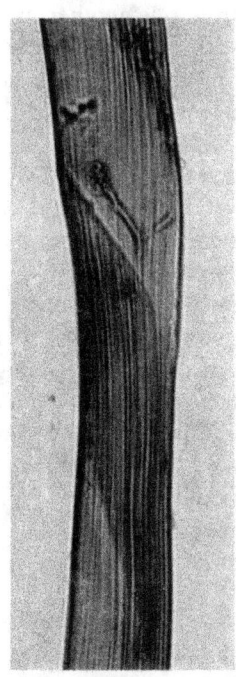

FIG. 64. Tussah silk, degummed (150x).

FIG. 65. Tussah silk, degummed (500x).

MINERAL FIBRES

ASBESTOS

THERE ARE two groups of minerals classified under the general term "asbestos," namely amphibole and serpentine, both of which occur in several forms.

Chrysolite, a variety of serpentine, is the most important form as a mineral fibre and is also the most commonly used in the making of asbestos products. It is found mainly in Canada, the United States, and Russia.

Asbestos is used in the manufacture of specialty papers, as a textile fibre, and in products where heat and chemical resistance are required. It can be dyed. Dyed asbestos cloth is used for draperies, curtains, rugs, etc. Although pure asbestos can be spun, it is normally blended with up to 15 per cent of other fibres for improved spinning properties.

Asbestos fibres are very fine and curly with smooth surfaces (Figs. 66 and 67). When examined under a microscope, they show a grouping or crowding together of many finer threads within what appears to be a single fibre. However, this "fibre" can be subdivided beyond the resolution of the optical microscope. Latest estimates on cross-sectioned diameters of the ultimate fibres vary from 214 to 285 Å.[1]

Recognition of asbestos in paper samples is by the fineness of the fibres, by the fact that they are in bundles, and that they remain unstained. It is also incombustible. On ignition of a sample, any organic matter present is burnt, but the asbestos only glows brightly without combustion taking place. When the ash is examined in a drop of water, the fibrillated asbestos fibres can be seen in ordinary transmitted light.[8]

GLASS

Glass fibres are now being used commercially to produce special papers, either from pure glass or from blends with other fibres. Because of dimensional stability, good electrical properties, and outstanding resistance to heat and chemicals, glass-fibre papers have many applications. For example, the better filter papers are made from glass fibres. The dimensional stability of cellulosic paper is greatly increased by adding a small percentage of glass fibres.

Glass fibres used in papermaking are very fine, varying from nine microns to less than one micron in diameter. They are also very pliable and silky.[12]

Microscopic examination[1] of glass fibres (washed clear of any lubricant or binder) reveals that they are perfectly smooth, having no visible structure on the surface (Fig. 68 and 69). The edges of a fibre are always parallel and lack any irregularities whatever. Even examination at the high magnifications possible with the electron microscope fails to reveal any roughness or irregularity.

Cross-sectional examination shows that the fibres are perfectly circular. Fibre diameters may vary considerably from the average. "Tramp" fibres, which are considerably coarser than the average, may appear in the fibre strands. An abnormality found in glass textile fibres is the occasional presence of tubular or hollow fibres which vary from a fraction of an inch to a few inches long. The characteristic fracture of glass fibres is a clean transverse or slightly diagonal break. They rarely or never split or break longitudinally.

Glass fibres have a high resistance to all acids (except hydrofluoric and hot phosphoric) and a moderately good resistance to alkalis. They are incombustible but soften and melt at very high temperatures.[1, 8]

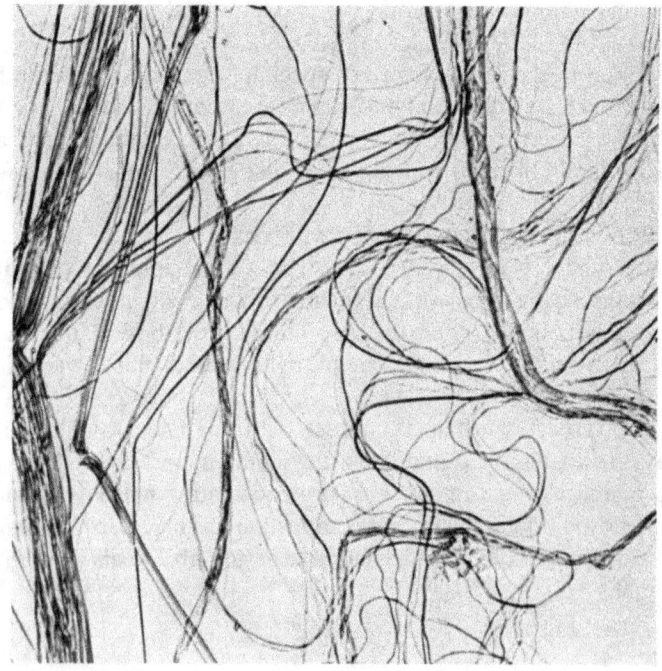

FIG. 66. Asbestos fibres (150x).

MINERAL FIBRES

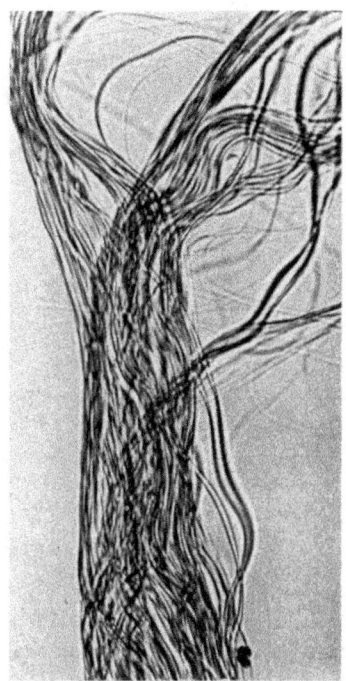

FIG. 67. Asbestos fibres (500x).

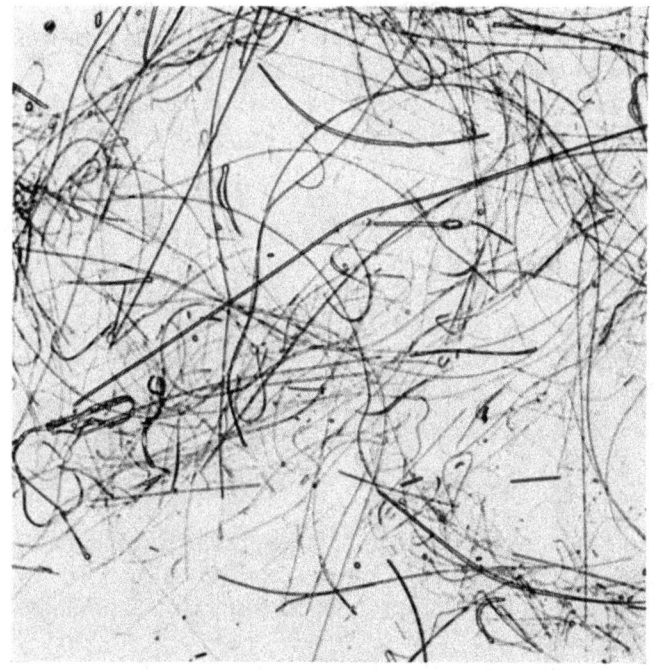

FIG. 68. Glass fibres (150x).

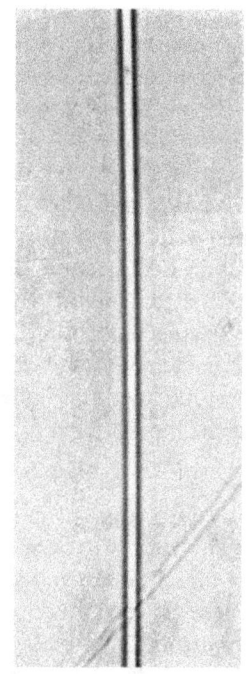

FIG. 69. Glass fibres (500x).

ORGANIC MAN-MADE FIBRES

RAYONS

BY DEFINITION, rayon is a generic name for man-made textile fibres and filaments composed of regenerated cellulose. The various types of rayon have been named after their commercial processes: viscose rayon, cuprammonium rayon, and nitrocellulose rayon. Nitrocellulose rayon is practically unknown nowadays as it has been replaced by the other rayons. The earliest rayons were commercially produced in Europe. They were known under the name of *artificial silk*. Often there is mention of acetate rayon, although this fibre is not produced from regenerated cellulose but instead from cellulose acetate. This is frequently classified with the rayons, but it is correctly defined as an acetate or cellulose acetate fibre or filament.

Rayons, like many other manufactured fibres, are produced in continuous filament as well as staple (short-length) forms. Except for the length, these two forms of fibres are similar in all respects.

Rayons and other man-made fibres are said to be "bright" if they appear glossy, and "dull" if they are not. Glossiness is reduced by adding a fine powder of titanium dioxide to the spinning solution. These fine particles scatter the light, making the fibres appear dull.

Rayons and other textile fibres are produced in different degrees of fineness (diameters), measured in deniers. The *denier*, a textile term, is the weight in grams of 9,000 metres of the filament or yarn. If 9,000 metres of a filament weigh 5 grams, the denier is 5.

VISCOSE RAYON

The viscose process was the last rayon process to be developed, but it is now by far the cheapest and most important. Because of the low cost, more than 80 per cent of all manufactured fibres are made by this process. Viscose rayon is produced from wood pulp and/or cotton linters.

Several types of viscose rayon are manufactured (Figs. 70 to 75). The conventional rayon is made in both filament and staple forms of varying lustre and denier.

It exhibits special features in the form of a number of striations parallel to the fibre axis. These striations are caused by the shrinking of the filament after it has left the spinneret. They may vary in a random manner and are more obvious in bright than in dull fibres. The striations are similar to those in celulose acetate fibres, however, viscose rayon is not soluble in acetone, while cellulose acetate is. Most viscose filament types have a skin which is denser than the core and is less permeable to dye solutions (see cuprammonium rayon).

The cross-sections of viscose fibres are valuable for identification. They may vary in size, according to denier, as well as in shape. The shape may vary from circular and irregularly oval to ribbon-like. The margins of the cross-sections are irregularly serrated. These serrations are numerous and sharp but not very deep.[1]

Viscose rayons were found to be the most suitable man-made fibres for papermaking and for the production of nonwoven fabrics which also are made on paper machines. For this purpose, the American Viscose Division produces a multicellular self-bonding fibre. This fibre has air pockets and its bonding degree is controlled by the number, size, and wall thickness of these pockets. The self-bonding properties make it possible to manufacture paper from 100 per cent man-made fibres without using special bonding agents.

Another rayon called "hybrid" is made for the same purpose. It fibrillates on mild mechanical beating in water and acquires self-bonding properties. Conventional rayons do not normally fibrillate on mechanical beating, but after special chemical treatment they fibrillate readily on beating in water. They are then used in papermaking.[12]

Viscose rayon (as well as cuprammonium rayon) swells in water from 25 to 45 per cent. For microscopic examination it is advisable to mount it in glycerine (refractive index, $r.i. = 1.4729$), colourless mineral oil ($r.i. = 1.46$), or monobromonaphthalene ($r.i. = 1.66$).[1]

Viscose rayon is attacked by hot dilute or cold concentrated mineral acids. It has high resistance to dilute alkalis. Strong solutions of alkali cause swelling. It is insoluble in most organic solvents,[10] including acetone.

Viscose fibres are markedly birefringent, more so than are acetate fibres. As with natural cellulose, Herzberg stain also reacts with regenerated cellulose and stains viscose fibres red to red-violet.

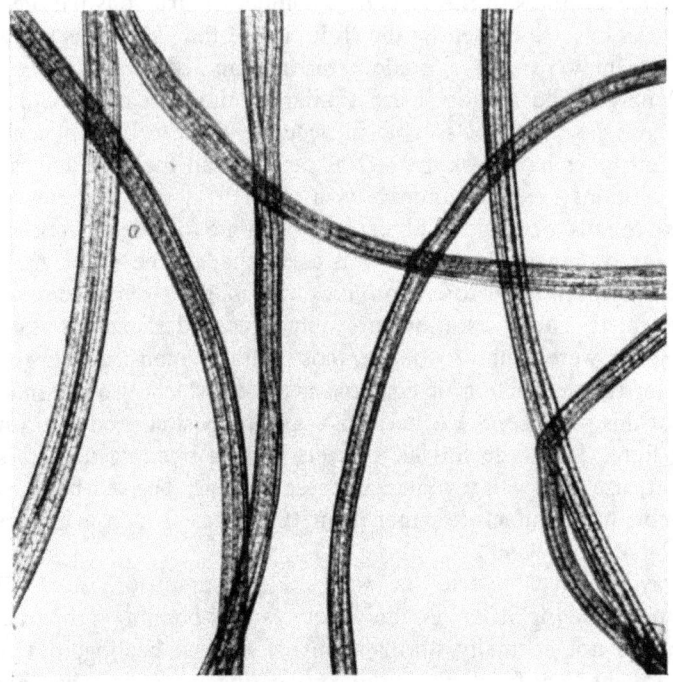

FIG. 70. Viscose rayon, dull (150x).

FIG. 71. Viscose rayon, bright (150x).

ORGANIC MAN-MADE FIBRES

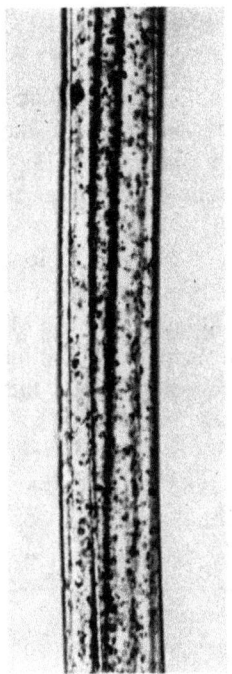

FIG. 72. Viscose rayon, dull (500x).

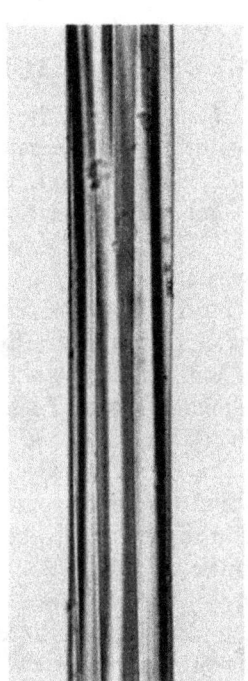

FIG. 73. Viscose rayon, bright (500x).

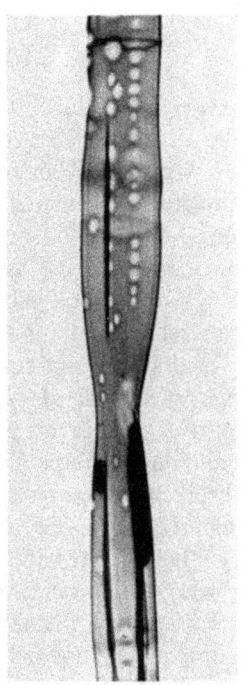

FIG. 74. Viscose rayon with "bubbles," self-bonding (150x).

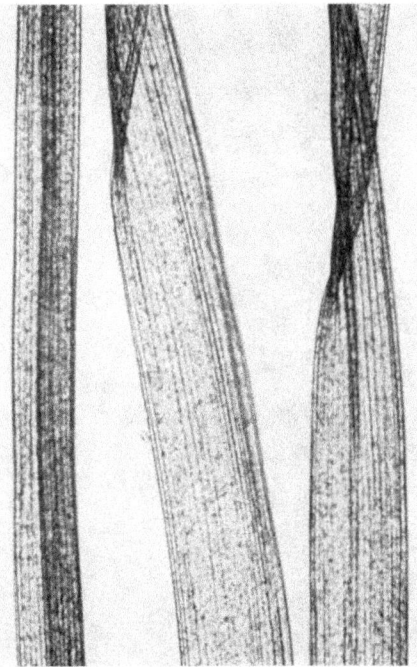

FIG. 75. Viscose rayon, flat (150x).

CUPRAMMONIUM RAYON

This rayon is produced in fine deniers and of various lustres: bright, dull, and semi-dull. It is also known as Bemberg rayon because it is a product of the American Bemberg Company. Like viscose rayon, cuprammonium rayon also swells in water. For microscopic examination it needs to be mounted in glycerine or colourless mineral oil.

Cotton linters are the usual raw material. They give rayon of a better colour and strength than wood cellulose. Under the microscope, the filaments appear fine and structureless, without striations or markings of any kind (Figs. 76 and 77). In cross-section they are circular or sometimes slightly oval, with a smooth contour which is very typical of cuprammonium rayon.[1] These features differentiate this rayon from viscose rayon and cellulose acetate fibres.

Cuprammonium fibres have no skin, the structure is also more open, and direct dyes are taken up more quickly than by most types of viscose which do have a skin. Consequently, cuprammonium rayon can also be differentiated from viscose rayon by a colour test with Brilliant Benzo Blue 6 BA (a 0.2 to 0.5 per cent solution in water) at room temperature. This is followed by rinsing in warm water. Cuprammonium rayon stains blue, whereas viscose fibres remain unstained.[4]

Cuprammonium rayon, as a fibre of regenerated cellulose, has many properties similar to those of viscose rayon. For example, it stains red to red-violet with Herzberg stain, has approximately the same resistance to acids, alkalis, and organic

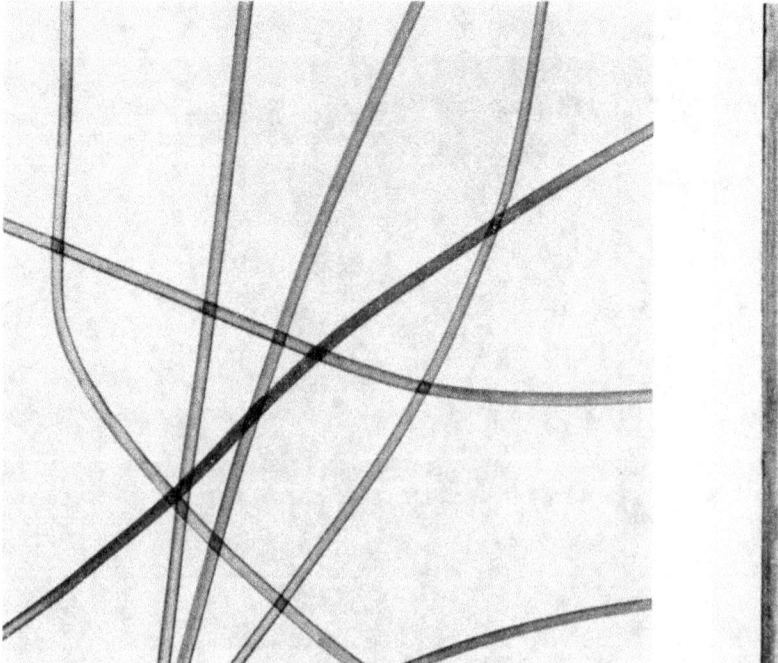

FIG. 76. Cuprammonium rayon (150x).

FIG. 77. Cuprammonium rayon (500x).

solvents, and is almost as birefringent as viscose rayon. Like conventional types of viscose, it is soluble in cuprammonium hydroxide.[8]

ACETATE

By definition, acetate fibres (also known as cellulose acetate or acetate rayon) are man-made textile fibres or filaments composed of cellulose acetate. They are manufactured from purified cotton or wood pulp cellulose.

Acetate fibres are produced in filament and staple forms of both bright and dull lustre and of various diameters (Figs. 78 to 81). Bright lustre fibres are clear and transparent. In dull lustre fibres, pigment particles of the delustring agent are visible as tiny specks which are distributed throughout the fibre substance. Acetate fibres swell less than 10 per cent in water.[1]

In microscopic, longitudinal view, striations similar to viscose rayon are present (but are more pronounced), though their occurrence is less frequent. These striations are visible as gently rounded, longitudinal ridges and valleys. The cross-sections of acetate fibres usually have a cloverleaf appearance with two, three, four, or occasionally more, large, smoothly rounded lobes. These are readily distinguishable from the serrated edges of cross-sections of the usual viscose rayons.[1]

Acetate fibres can usually be identified by microscopic examination. Confirmatory tests include solubility in cold acetone. For this test the fibres are mounted dry between a slide and a coverglass and are observed under the microscope, while acetone is added to the slide next to the edge of the coverglass. In this manner the

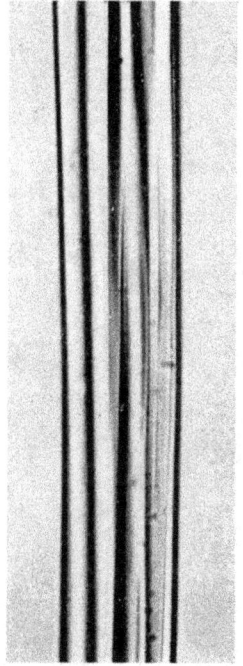

FIG. 78. Acetate, bright (500x).

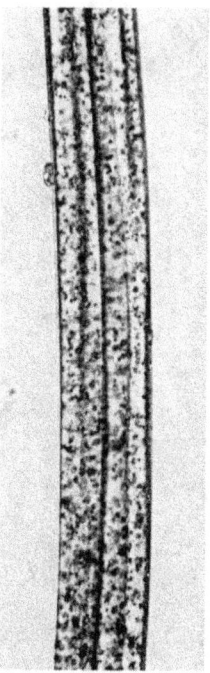

FIG. 79. Acetate, dull (500x).

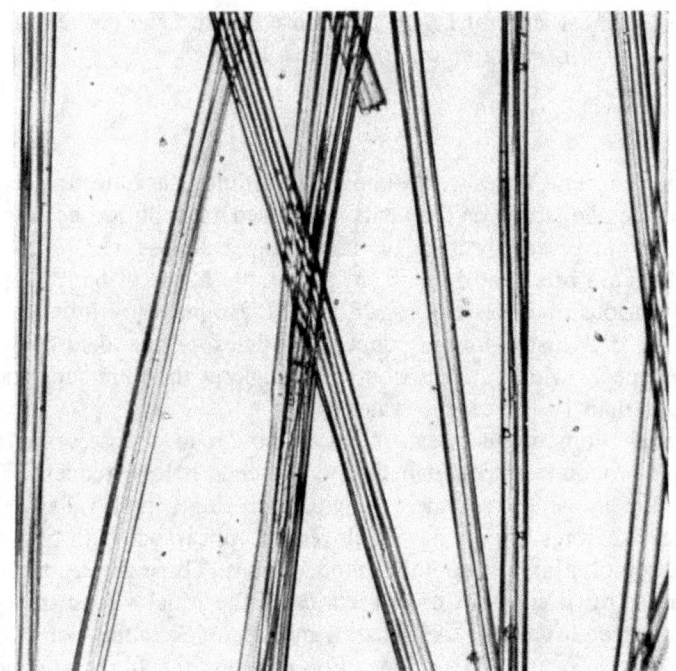

FIG. 80. Acetate, bright (150x).

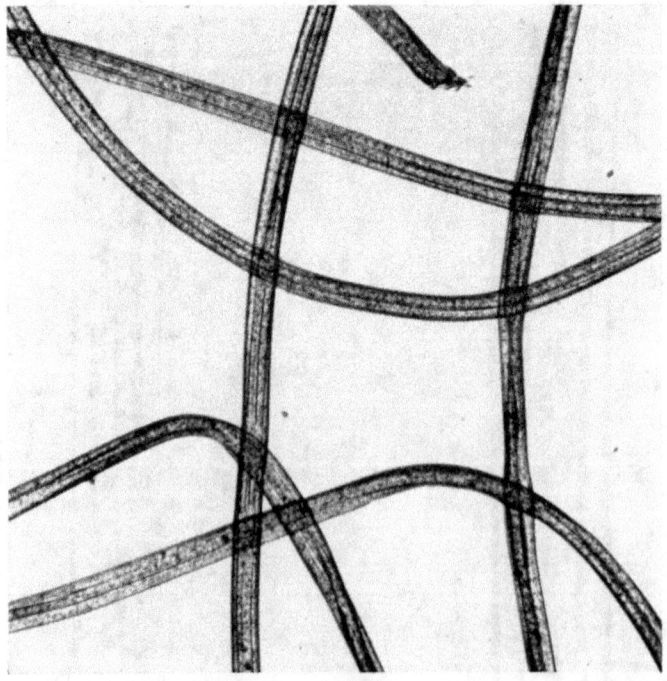

FIG. 81. Acetate, dull (150x).

fibres may be seen to dissolve the instant the acetone touches them. When a dulled fibre is dissolved in this way, the dulling material remains as a residue on the slide.[4]

The dyeing properties of cellulose acetate are very different from those of other fibres. In general, acetate is not dyed by dyes which dye cellulose and protein fibres, and those special dyes which will dye cellulose acetate will not dye the other fibres.[1]

The birefringence of these fibres is very low, so that they appear fairly dark in polarized light at total extinction. Unlike the results with viscose and cuprammonium rayons, Herzberg stain develops a yellow colour with acetate fibres, since they are not of cellulose but of cellulose acetate.

Cellulose Triacetate

The difference in appearance between triacetate and normal (secondary) acetate is insufficient for positive identification. However, methylene dichloride dissolves triacetate but only swells secondary acetate; 80 per cent acetone swells triacetate but dissolves secondary acetate.[9]

NYLON

Nylon was the first synthetic textile fibre to be produced. It is made from a polyamide condensation polymer. Prior to adoption of the present name, nylon was known as "Fibre 66."

Commercial production of nylon was started by the Du Pont Company in 1938. It is produced in filament and staple forms of various diameters. There are both bright and dull fibres. Dull fibres show the same features as bright ones, except that the particles of the delustrant are present both on the surface and within the fibres themselves.

In longitudinal view, the sides of nylon fibres are completely parallel all along their length (Figs. 82 to 85). The surfaces are smooth and free from irregularities. The fibres are almost perfectly round in cross-section. Because of their nearly perfect cylindrical shape and smooth and lustrous surfaces, nylon fibres resemble glass rods when examined under a microscope.

Nylon and Dacron fibres are quite similar in their microscopic appearance. Nylon is specifically differentiated from Dacron and all other present synthetic fibres by its high solubility in 18.5 per cent solution of hydrochloric acid at room temperature.[4] Nylon fibres dissolve the instant the acid touches them, though Dacron fibres remain unaffected by it.

Another important method of identification of nylon is the burning test. This test is carried out by burning a tiny bundle of fibres over a small flame. Nylon fibres do not flash burn but melt and develop a characteristic amide odour of burning hair. The molten portion from the burning fibres forms a glossy bead of extraordinary hardness. Nylon fibres are also recognized by their high birefringence. As a result they have bands of polarization colours along the fibre edges.[4]

Nylon is not readily affected by alkalis, even hot solutions, but it disintegrates in cold concentrated hydrochloric, sulphuric, and nitric acids and in a boiling solution of 5 per cent hydrochloric acid. It is also soluble in 90 per cent phenol and concentrated formic acid. It is not soluble in acetone. Nylon, like other synthetic fibres, swells very little (2 per cent) in water.[10]

A small amount of nylon is used by the paper industry. Because in sheet formation long fibres tend to get entangled and form clumps, it is necessary to cut the nylon filaments into lengths of between ⅛ and 1 inch long. The variation in length depends on filament diameter. The fibres are not self-bonding. This does not change on beating as the fibres do not fibrillate. Consequently, nylon fibres must be blended with wood pulp and/or special bonding agents if satisfactory sheet formation is to be achieved. The bonding agents include (1) aqueous polymer dispersions, which are available in a wide range of compositions, and (2) fibrous condensation polymers, such as polyamide (nylon) fibrids.[12]

Dispersion binders are applied by precipitation on the fibres before the web is formed or by saturation after web formation. On the other hand the fibrous binders (fibrids) must be melted to bring about bonding of the fibres. A commercial nylon paper called Syntosil is reported as being used for maps and currency.[12]

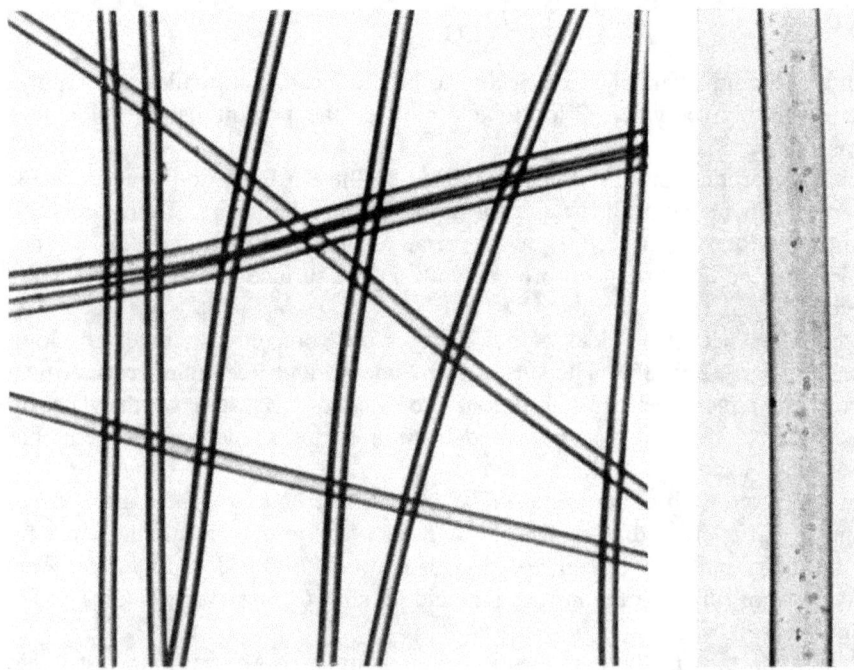

FIG. 82. Nylon, bright (150x).

FIG. 83. Nylon, almost bright (500x).

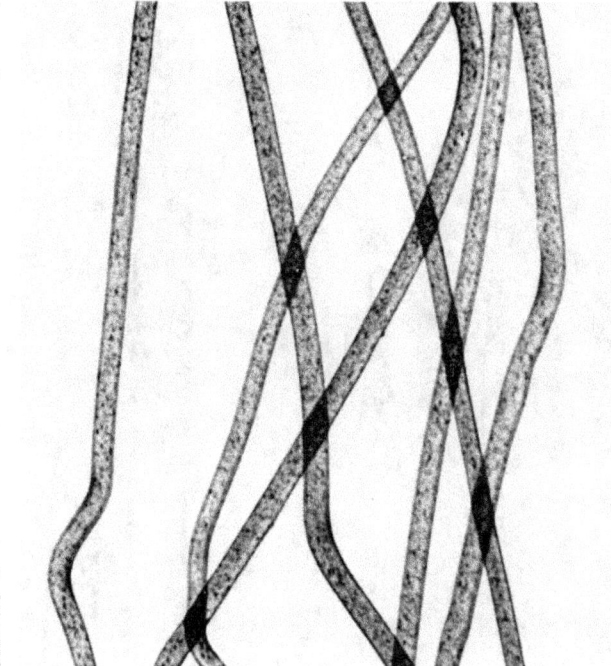
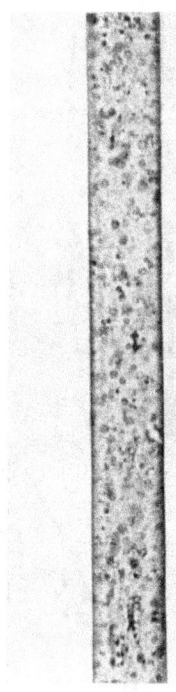

FIG. 84. Nylon, dull (150x).	FIG. 85. Nylon, dull (500x).

DACRON—TERYLENE

This polyester fibre is known as Dacron in the United States and as Terylene in England. It is manufactured from a polymer derived from ethylene glycol and terephthalic acid. Dacron is the product of the Du Pont Company.

Dacron filaments and staple fibres (Figs. 86 and 87) have microscopic features similar to those of nylon.[1] They are transparent, very uniform in diameter, circular in cross-section, and have more or less pitted surfaces. Even a bright Dacron contains a small amount of delustrant, and its particles appear on the surface as pits. These pits or dulling particles are in longitudinal rows.[4]

Dacron has a good resistance to most weak acids and alkalis and a moderate resistance to strong acids and alkalis at room temperature. It is unaffected by cold water, alcohols, acetone, and some other organic solvents. Dacron is dissolved by concentrated sulphuric acid and some phenolic compounds such as m-cresol.[1, 8] Dacron is differentiated from nylon by its resistance to hydrochloric acid (see Nylon) and by appreciably lower birefringence.

Dacron fibres melt and char on ignition and eventually burn with a yellow, blue-edged flame, producing a heavy, aromatic smoke. Dacron is not stained by most water-soluble dyes but is dyed by some acetate type dyes.[4]

In order to obtain good sheet formation in papermaking Dacron filaments are cut into lengths from ¼ to ¾ inches. As they are not self-bonding, they are used in blends with beaten wood pulp and/or special bonding agents[12] (see Nylon).

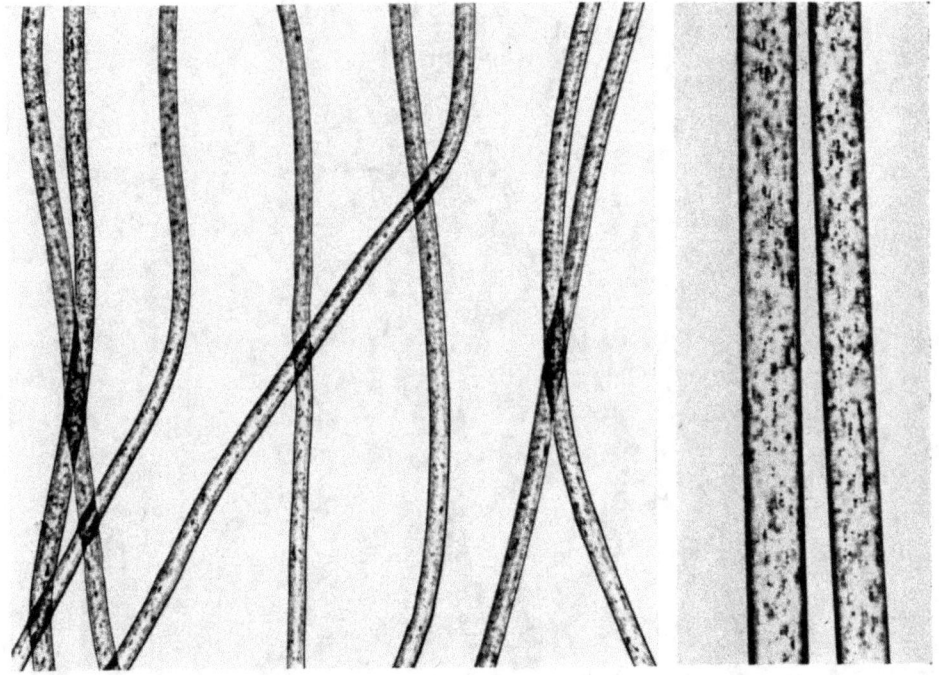

FIG. 86. Dacron (150x). FIG. 87. Dacron (500x).

ORLON

Orlon, an acrylic fibre, is a polymer of acrylonitrile. It has been commercially manufactured, in both filament and staple forms, by Du Pont since 1950.

The general appearance of Orlon is to some extent similar to Acrilan, except that the surface of Orlon is irregularly striated to a varying degree. The fibre has a "skin" which is denser than its core (Figs. 88, 89, 90). In cross-section Orlon has a dumb-bell or dog-bone shape. In longitudinal view the fibre shows internal fibrillation and the thinner portion of the dumb-bell creates the impression of a false lumen.[4] Orlon has a low birefringence.

In general, Orlon does not swell appreciably in water, it has a good resistance to mineral acids, weak alkalis, and most common organic solvents.[10] It is dissolved by concentrated sulphuric and nitric acids.

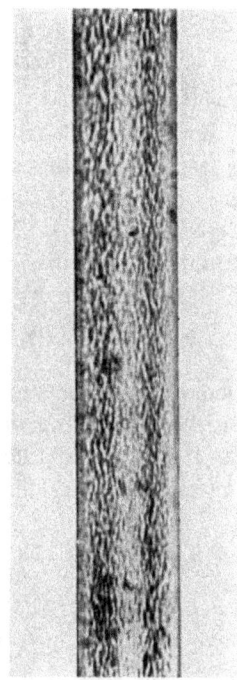

FIG. 88. Orlon. Note the striated surface. (500x).

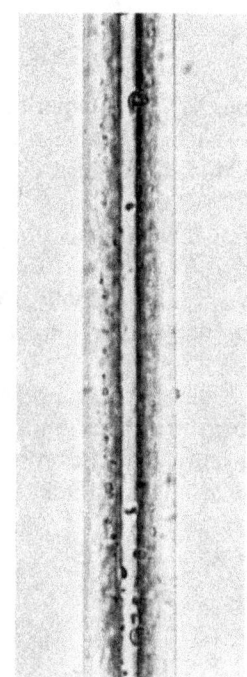

FIG. 89. Orlon. Note the groove along the fibre. (150x).

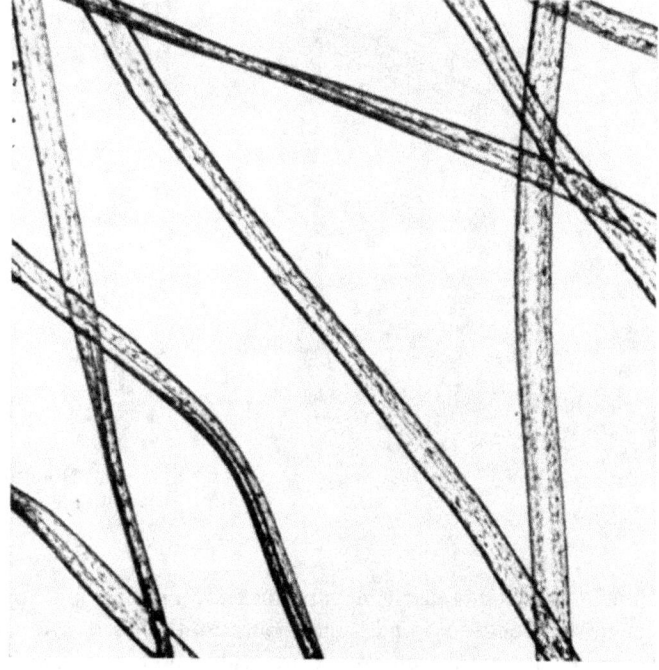

FIG. 90. Orlon (150x).

ACRILAN

Acrilan, an acrylic fibre, is a copolymer of acrylonitrile with vinyl derivatives. It is produced in a wide range of diameters (deniers) and in both filament and staple forms. It is a product of Chemstrand Corporation and was first produced in 1952.

The Acrilan fibre has a single groove along its length which gives it a dumb-bell shaped cross-section (Figs. 91 and 92). Because of this groove the fibre seems to have a lumen when viewed longitudinally. Acrilan has typical short, but distinct, striations which run parallel to the long axis of the fibre.[8] It has a very low birefringence.

Acrilan swells by about 5 per cent in water. It is dissolved by concentrated sulphuric acid and fuming nitric acid at room temperature.[4] On the other hand, it has a good resistance to weak as well as most other strong acids. It has a moderate resistance to weak and strong cold alkalis. It is resistant to common organic solvents[1] such as acetone, etc.

FIG. 91. Acrilan (150x).

FIG. 92. Acrilan. Note the short striation. (500x).

Papermaking

Acrylic fibres (Orlon and Acrilan) are used in papermaking. For this purpose they have to be cut into lengths of between ⅛ and ¼ inch. Because the fibres are not normally self-bonding it is necessary to use an admixture with wood pulp and/or special bonding agents (see Nylon). However, there is a special process, involving beating in water, by which acrylic fibres can be fibrillated so that self-bonding occurs. In this way it is possible to produce paper from pure acrilic fibre.[12]

VINYON N—DYNEL

Vinyon, a vinyl fibre, was originally manufactured from a copolymerized mixture of vinyl chloride and vinyl acetate. This form, however, was sensitive to heat over 65°C. The later types of Vinyon, called Vinyon N in continuous form and Dynel in staple form, are manufactured from a copolymer of vinyl chloride and acrylonitrile, and are less sensitive to heat. Their fibres are also of identical appearance.

In longitudinal view the fibres appear ribbon-like in shape and often are twisted or even folded double (Figs. 93, 94, 95). The ribbon is often thicker on one side than on the other. It shows striations and some internal fibrillation. The fibre surface is somewhat coarse.[4] The single groove on one side of the fibre creates the impression of a false lumen somewhat similar to that seen in Orlon. The cross-section is slender and irregularly dumb-bell shaped. Vinyon N and Dynel have low birefringence.

These fibres do not swell appreciably in water. They have a good resistance to both strong acids and alkalis. They are not affected by common organic solvents.[10] However, they swell in concentrated nitric acid and are dissolved by concentrated sulphuric acid, boiling concentrated acetone, and warm 90 per cent phenol.[8]

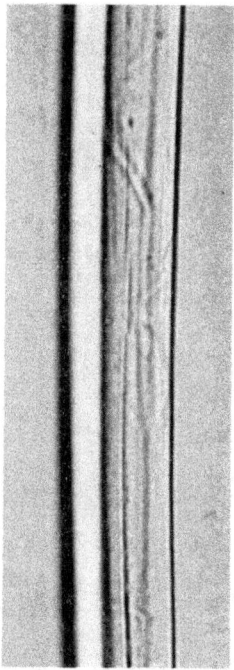

FIG. 93. Dynel. Note the striations and the difference in thickness of the sides. (500x).

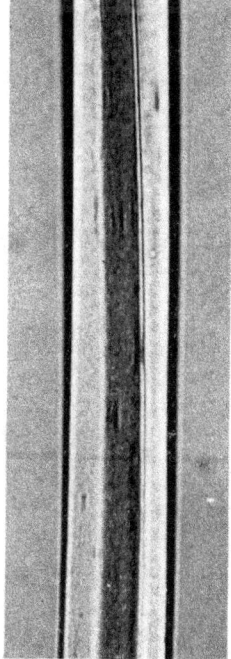

FIG. 94. Dynel (500x).

FIG. 95. Dynel (150x).

REFERENCES

1. Mauersberger, H. R. Mathew's Textile Fibres, 6th ed.; New York: John Wiley and Sons, 1954. 1283 pp.
2. Isenberg, I. H. Pulp and Paper Microscopy, 3rd ed.; Appleton, Wisconsin: Institute of Paper Chemistry, 1958. 333 pp.
3. Armitage, F. D. An Atlas of the Commoner Paper Making Fibres, London: Guildhall. 172 pp.
4. Heyn, A. N. J. Fiber Microscopy, New York: Interscience Publishers, 1954. 407 pp.
5. Identification of Textile Materials, Textile Institute, 10 Blackfriars Street, Manchester, 3, 1951. 94 pp.
6. Species Identification of Nonwoody Vegetable Fibres, T10 m–47, tentative standard of October 1947, Technical Association of Pulp and Paper Industry, 122 East 42nd St., New York 17, N.Y. 8 pp.
7. Wangaard, F. F. Identification of Fibres Other Than Wood Used in Paper Manufacture, Paper Industry, October 1937, pp. 777–84, 794.
8. Harris, Milton. Handbook of Textile Fibres, 1st ed.; Washington: Harris Research Laboratories, 1954. 356 pp.
9. Montcrieff, R. W. Man-Made Fibres, New York: John Wiley and Sons, 1963. 742 pp.
10. Cook, J. G. Handbook of Textile Fibres, Watford, Herts, England: Merrow, 1959. 422 pp.
11. Werner von Berger *and* Walter Krauss. Textile Fibre Atlas, New York: American Wool Handbook Company, 1942. 34 pp. and 25 plates.
12. Battista, O. A. Synthetic Fibers in Papermaking, New York: Interscience Publishers, 1964. 340 pp.

APPENDIXES

I. SOLUBILITY

Solvents	Wool	Silk, cultivated	Tussah silk	Cotton	Other vegetable fibres	Viscose rayon
Acetone, concentrated						
Acetone, 80 per cent						
Calcium thiocyanate, concentrated, hot						×
Hydrochloric acid concentrated, at 30–40°C		×				
Methylene chloride						
Nitric acid concentrated		dis-integrates				
Phenol 90 per cent						
Caustic soda 5 per cent, boiling	×	×	partly ×			
Cuprammonium hydroxide		×	×	×	not all ×	×
Sulphuric acid concentrated		×	×	×	slowly ×	×

NOTE: × indicates solubility. [1, 4, 5, 8, 10] See also notes to Appendix II.

OF FIBRES

Cuprammonium rayon	Acetate (secondary)	Triacetate	Nylon	Vinyon N, Dynel	Dacron	Orlon	Acrilan
				hot ×			
	×	swells					
×							
	×	×	×				
	swells	×					
	×	×	×	swells		×	×
	×	×	×	warm ×	warm ×		warm ×
	saponifies slowly						
×	×						
×	×	×	×	×	×	×	×

II. A SCHEME OF FIBRE ANALYSIS BY SOLVENT ACTION

This scheme[4] was suggested by William R. Wilson. If it is used with a blend of fibres as listed below, each will be dissolved in turn, Dacron and the primary wall of cotton being left behind.

Fibre	Solvent
Cellulose acetate	Glacial acetic acid
Vinyon HH (not described here)	Chloroform
Dynel and Vinyon N	Dimethylformamide
Nylon	90 per cent Phenol
Viscose rayon	Cuprammonium hydroxide
Cuprammonium rayon	Cuprammonium hydroxide
Cotton	Cuprammonium hydroxide (partially dissolved, balloons)
Wool	20 per cent Bleach (NaOCl)
Vicara (not described here)	20 per cent Bleach (NaOCl)
Orlon	Dimethylformamide if stirred in a beaker at 55 to 60°C but not on a slide.
Acrilan	Concentrated nitric acid
Dacron	—

Notes: For such tests, dry fibres on a micro slide are covered with a coverglass. The solvent is delivered to the edge of the coverglass, and the reaction is observed under the microscope.

The use of a single reagent for the identification of an unknown fibre is not recommended. Solubility tests should rather serve as a check for the microscopic identification.

The chemical processes to which fibres have previously been subjected may modify the solvent reactions shown above.

One of the most reliable means of identification is by comparison of the fibres under investigation with fibre samples of known origin and solvent reaction.

III. COLOUR REACTIONS OF FIBRES TO SELECTIVE STAINING

THREE SELECTIVE STAINS are here noted because they appear to be most suitable for identification of fibres, and because they also differentiate some of the natural fibres. In using these stains it is always advisable to make comparisons with authentic samples. It is important to follow the dye manufacturer's recommendations for the use of the selective stains.

Shirlastain A

This stain is a liquid ready for use and is supplied with instructions and colour chart by Shirley Developments Limited, 52–56 Market Street, Manchester 1, England. The fibres are stained at room temperature for one minute, rinsed in cold tap water, and compared with the colour chart.

Colotex B

Colotex B is a dye stain made by Union Chemical Company, New York, and distributed by Neuberg Chemical Corporation, 441 Lexington Avenue, New York. The fibres are immersed for 3 to 5 minutes in the stain, washed until the water is free from colour, and passed through water containing a few drops of ammonia. Finally, the fibres are rinsed again in fresh water and dried.[1]

Neocarmine W

This stain may be purchased from F. F. Kraus Company, 136 Liberty Street, New York 6. It is supplied with a description of the staining procedure and a colour chart. This stain is similar to Colotex B and so is the staining procedure.

Fibre	Shirlastain A[5]	Colotex B[1]	Neocarmine W[1]
Cotton, raw	Pale dusty purple	Dull lavender	Light blue
Cotton, mercerized	Mauve	Blue violet	Deep blue
Cotton, bleached	Purple	Dull violet	Deep blue
Kapok (Java)	Golden yellow	Greenish yellow	Greenish yellow
Flax, raw	Brownish purple	Bluish mauve	Dull deep blue
Flax, bleached	Violet blue		
Hemp	Dark purplish grey	Dull light brown	Violet blue with red dots
Ramie	Lavender		Blue violet
Jute	Golden brown	Light reddish brown	Olive brown
Sisal	Golden brown		Greenish yellow
Wool	Cold: bright yellow / At boil: copper brown	Deep maize	Yellow
Silk, raw	Dark brown		
Silk, degummed	Golden brown	Reddish tan	Dull gold
Silk, Tussah	Pale chestnut brown	Golden yellow	Green
Viscose rayon	Bright pink	Lilac	Red violet
Cuprammonium rayon	Bright blue	Reddish navy blue	Deep blue
Cellulose acetate	Bright greenish yellow	Lemon yellow	Greenish yellow
Nylon	Cold: cream to yellow / Hot: copper brown	Dull reddish yellow	Greenish
Dacron	Unstained		
Orlon	Cold: unstained / At boil: salmon pink		
Vinyon	Unstained	Unstained	Pale yellow
Coir		Dull yellow brown	Light brown

IV. IDENTIFICATION KEY TO NONWOODY AND MAN-MADE FIBRES

THE KEY is used for the identification of unknown fibres, in conjunction with microscopic examination, and is read beginning with feature 1. Depending on the characteristics of the feature being observed, the examiner has a choice that will lead either to a conclusive identification (e.g., 1.a. leads to "wool or animal hairs") or to a further feature number (e.g., 1.b. leads to feature 2). The process is repeated through successive features until such time as a positive identification is achieved.

Similarly, if for example, the examination of an unknown sample shows that the fibres do not have surface scales (1.b.), it must be examined for the characteristics listed under feature 2. It is then observed that the fibres have a cellular structure (2.b.) and that the occurrence of other cellular elements is rare [2.b.(i)]. This leads on to feature 3. As the fibres are found to be long with irregularly spaced convolutions (3.a.) this leads to feature 4. Observation of the characteristics listed under this feature shows that the fibres are ribbon-shaped, have large lumens, and thickened walls (4.a.), leading to a positive identification of cotton.

1.a. Fibres have surface scales — Wool and animal hairs
  b. Fibres do not have surface scales — 2
2.a. Curly fibres composed of aggregates of fine fibres, showing no structural details. Incombustible — Asbestos
  b. Fibres exhibiting a cellular structure, i.e., lumen, cell wall and natural ends:
    (i) fibres with rare or no other cellular elements present   (bast, leaf, and seed hair fibres) — 3
    (ii) fibres plus numerous other cellular elements present, including epidermal and barrel-shaped parenchyma cells and vessel segments (grass pulp) — 8

c. Fibres do not show cellular structure but appear as continuous filaments with constant features repeated along their length:
      (*i*) width varies along their length — 7
      (*ii*) width uniform along their length, bright or delustered — 13
3.a. Long fibres with convolutions, irregularly spaced — 4
  b. Fibres do not normally have convolutions. Fibre ends tapering — 5
4.a. Ribbon-shaped fibres with large lumens and thickened edges — Cotton
  b. Fibres of almost cylindrical shape, faint or occasionally pronounced convolutions, fine lumens and smooth surfaces — Mercerized cotton
5.a. Ribbon-like fibres having thin cuticles, very narrow and indistinct lumens. Fibres 6 to 20 mm long — Paper mulberry
  b. Flat, short fibres with diagonal striations and pits. Fibres 0.3 to 1.0 mm long — Coir
  c. Width of fibres varies throughout their length. Fibre walls are thick, show striations and cross-markings
      (*i*) Lumen not wide, fibres 60 to 250 mm long — Ramie
      (*ii*) Lumen wide, fibres 5 to 55 mm long — Hemp
      (*iii*) Central segment of fibres wider, 2.4 to 3.6 mm long — Mitsumata
  d. Width of fibres relatively uniform throughout their length — 6
6.a. Fibre lumen varying in width throughout its length. Walls show cross-markings.
      (*i*) Fibres with surrounding lignin layer, ends blunt,
        2 to 6 mm long — Kenaf
        4 to 12 mm long — Sunn
      (*ii*) Fibres with surrounding lignin layer, ends slender and pointed,
        1.5 to 5 mm long — Jute
      (*iii*) Fibres with no surrounding lignin layer, ends thick and usually blunt, some pointed.
        1.5 to 4 mm long — Sisal
  b. Fibre lumen of uniform width and very narrow, almost line-like, often indistinct. Walls are thick, ends pointed.
      (*i*) Cross-markings I, X, V shaped. Fibres 5 to 55 mm long — Flax
      (*ii*) Very fine fibres, 1.7 to 10 mm long — Pineapple
      (*iii*) Fine fibres, 2 to 6 mm long — Caroa
      (*iv*) Fibres 2 to 15 mm long — N.Z. flax
  c. Fibre lumen of uniform width and very wide. Walls are thin to very thin.

(i) Smooth, structureless, surface. Forms air bubbles in lumen when immersed in water. Fibres 7.5 to 31 mm long — Kapok

(ii) Walls have many cross-markings. Fibre ends pointed. Fibres 2 to 12 mm long — Manila hemp

(iii) Walls have no special markings. Fibre ends blunt. Fibres 1.3 to 6 mm long — Mauritius hemp

7.a. Structureless, smooth filaments, often in pairs, with constrictions, swellings and lumps along their length. Gelatinise and dissolve in boiling 5 per cent caustic soda — Cultivated silk

b. Striated, ribbon-like filaments, with obliquely flattened cross-markings. Disintegrate in boiling 5 per cent caustic soda, but do not dissolve completely — Tussah silk (wild silk)

8.a. Epidermal cells abundant, their margins distinctly serrated — 9

b. Epidermal cells infrequent, their margins vary — 11

9.a. Barrel-shaped parenchyma cells infrequent and of maximum width of 20.5 microns. Epidermal cells less than 14 microns wide. Erect trichomes hooked at their apices — Esparto

b. Barrel-shaped parenchyma cells abundant — 10

10.a. Fibres very fine (5.1 to 13.6 microns wide). Epidermal cells with many pits and more than 14 microns wide. Trichomes sparse, erect and not hooked. Stomata of epidermis accompanied by dentate guard cells — Rice

b. Fibres 6.8 to 23.8 microns wide. Epidermal cells rarely pitted. Trichomes rare, but erect. Guard cells of stomata smooth-walled — Wheat, Oat, Barley, Rye

11.a. Fibres rather long—up to 4.3 mm, averaging 2.7 mm. Also present are thin-walled, wide (up to 40 microns) and ribbon-shaped fibres — Bamboo

b. Fibres up to 2.9 mm long, accompanied by broad, not ribbon-shaped fibres of maximum width 34 microns — 12

12.a. Vessel segments up to 1.35 mm long, parenchyma cells up to 0.85 mm long, maximum fibre width 34 microns — Sugar cane

b. Vessel segments up to 0.60 mm long, parenchyma cells up to 0.32 mm long, maximum fibre width 24 microns — Corn

13.a. Fibres with several longitudinal surface grooves — 14

b. Fibres with a single groove along its length which creates the impression that a lumen is present — 15

c. Fibres without grooves, perfectly cylindrical (or nearly so) and without surface irregularities — 16

14.a. Stains red to red violet with Herzberg stain, insoluble in acetone — Viscose rayon

b. Stains yellow with Herzberg stain, soluble in acetone. — Cellulose acetate

15. a. Fibre surface is irregularly striated to a varying degree. Insoluble in both hot and cold acetone — Orlon
   b. Distinct, and typical, short striations along the fibre axis. Insoluble in both hot and cold acetone. Soluble in warm 90 per cent phenol — Acrilan
   c. Ribbon-like in shape, thicker on one side than other, shows striations and internal fibrillation. Soluble in hot acetone and warm 90 per cent phenol — Dynel / Vinyon N
16. a. Stains dark yellow with Herzberg stain; soluble in 18.5 per cent hydrochloric acid — Nylon
   b. Does not stain with Herzberg stain; insoluble in hydrochloric acid — Dacron
   c. Does not stain. Soluble in hydrofluoric acid — Glass
   d. Stains red to red violet with Herzberg stain, insoluble in acetone. The fibre is not perfectly cylindrical — Cuprammonium rayon

INDEX

Numbers in bold type indicate illustrations.

DENIER, definition of, 90
Distinguishing of pulps: of hardwood, 29; of softwood, 29

FIBRE IDENTIFICATION, 4, 44; of nonwoody origin, 41; key to, 47, 111; of wood, 4, 10, 28
Fibres
  hardwood, *see* Vessel elements
  man-made, 46
    acetate, 95, **95**, **96**
    acrilan, 102, **102**
    cellulose acetate, 95
    cellulose triacetate, 97
    Dacron, 99, **100**
    Dynel, 103, **103**, **104**
    glass, 87, **89**
    nylon, 97, **98**, **99**
    Orlon, 100, **101**
    rayon, 90
      acetate, 95, **95**
      cuprammonium, 94, **94**
      nitrocellulose, 90
      viscose, 90, **92**, **93**
    terylene, 99
    Vinyon N, 103
  nonwoody, natural, 44
    abaca, 65
    asbestos, 87, **88**, **89**
    bamboo, 80, **80**
    barley, 74, **75**
    caroa, 69, **70**
    cereal straws, 74
    coir, 53, **54**
    corn, 77, **79**
    cotton, 48, **50**, **51**
      mercerized, 49, **51**
    esparto, 77, **79**
    flax, 55, **56**
      New Zealand, 68, **68**
    hairs, 82, **83**
    hemp, 56, **57**
      Ambari, 61
      Benares, 60
      Gambo, 61
      Manila, 65, **66**
      Mauritius, 70, **71**
      Sunn, 60, **60**
    Java kapok, 52, **52**, **53**
    jute, 58, **58**
    kenaf, 61, **61**
    kozo, 62
    kodzu, 62
    mitsumata, 63, **63**, **64**
    oat, 74, **75**
    paper mulberry, 62, **62**
    pineapple, 68, **69**
    ramie, 59, **59**
    rice, 76, **78**
    rye, 74, **76**
    silk, 84, **84**
      Tussah, 85, **86**
    sisal, 66, **67**
    straws, 72
    sugar cane bagasse, 76, **78**
    wheat, 72, **73**, **74**, **74**
    wool, 82, **83**
  softwood, 29
    cedars
      western red, 31, **36**
      yellow, 31, **36**
    firs, 31
      Douglas, 31, **35**
      balsam, **16**, 31, **36**
      true, 31
    hemlock, **16**, 31
      eastern, **35**
      western, **35**

larches, 31
  eastern, **35**
  tamarack, **35**
  western, **35**
pine
  jack, **14**, **15**, 30, **33**
  lodgepole, 30, **33**
  red, 30, **32**
  pitch, **33**
  ponderosa, 30, 31, **33**
  Scots, 30, **32**
  southern, 30
  white, eastern, **15**, 30, **32**
  white, western, 30, **32**
spruces, 31
  black, **34**
  Engelmann, **34**
  red, **15**, **34**
  Sitka, **34**
  white, **34**

HARDWOODS, 5, 8; diagnostic features of, 17; illustrations of, **19**; key of, 7, 12; key card of, 26; list of, 24

PULPS, 4, 10; hardwood, 36, **38**; softwood, 29, **32**; mechanical, 29; groundwood, 29; distinguishing of, 29; staining of, 29

SOFTWOODS, 5, 7; diagnostic features of, 12; illustrations of, **13**; key of, 7, 12; key card of, 26; list of, 24
Solubility tests: of fibres, 46; table of, 106, 108
Stains, selective, 46

Alexander, 29
Colotex B, 109
Korn, 29
Neocarmine W, 109
Selleger, 29
Shirlastain A, 109
tables of, 109

VESSEL ELEMENTS, 36
  of alder, red, **19**, **40**
  ash, white, **38**, **39**
  aspen, trembling, **39**
  basswood, **20**, **39**
  beech, **39**
  birch, **40**
  elm, white, **21**, **39**
  gum, red, **40**
  hickory, shagbark, **39**
  maple, sugar, **40**
  oak, red, **23**, **38**
  oak, white, **20**, **38**
  poplar, yellow, **40**
  tupelo, **40**
  willow, **39**

WOOD
  examination of, 7
    macroscopic, 7
    microscopic, 8
  identification of, 1, 5
    ability of, 6
    by genera, 6
    key to, *see* Hardwoods and Softwoods
    by species, 6

INDEX TO SCIENTIFIC NAMES OF TREES AND OTHER PLANTS

Abies amabilis, 25
  balsamea **14**, **16**, 25, **36**
  grandis, 25
  lasiocarpa, 25
Acer spp., 25
  saccharum, **24**, **40**
Agave sisalina, 66
Alnus rubra, **19**, **24**, 25, **40**
Ananas comosus, 68
Avena sativa, 74

Bambusa spp., 80
Betula spp., 25, **40**
Boehmeria nivea, 59
Broussonetia papyrifera, 62

Cannabis sativa, 56
Carya spp., 25
  ovata, **21**, **39**
Ceiba petandra, 52
Chamaecyparis nootkatensis, **16**, 25, **36**

Cocos nucifera, 53
Corchorus capsularis, 58
Crotalaria juncea, 60

Dendrocalamus arundinacea, 81

Edgeworthia papyrifera, 63

Fagus grandifolia, 25, **39**
Fraxinus spp., 25
  americana, **21**, **23**, **38**, **39**
Furcraea gigantea, 70

Gossypium spp., 48
  hirsutum, 48

Hibiscus cannabinus, 61
Hordeum spp., 74

Larix laricina, 25, **35**
  occidentalis, **13**, 25, **35**

INDEX

Linum usitatissimum, 55
Liquidambar styraciflua, **19, 23,** 25, **40**
Liriodendron tulipifera, **20, 21,** 25, **40**

Musa textilis, 65

Neoglazovia variegata, 69
Nyssa spp., 25, **40**

Oryza sativa, 76

Phormium tenax, 68
Picea engelmannii, 25, **34**
    glauca, **13,** 25, **34**
    mariana, 25, **34**
    rubens, **14, 15,** 25, **34**
    sitchensis, 25, **34**
Pinus spp., **14, 15**
    banksiana, **14, 15,** 25, **33**
    contorta, **17,** 25, **33**
    echinata, 25
    elliottii, 25
    monticola, 24, **32**
    palustris, 25
    ponderosa, 25, **33**
    resinosa, 24, **32**
    rigida, 25, **33**
    serotina, 25
    strobus, **15,** 24, **32**
    sylvestris, 24, **32**
    taeda, 25
Populus spp., 25
    tremuloides, **20, 39**
Pseudotsuga menziessii, **16, 17,** 25, **35**

Quercus spp., 25, **38**
    alba, **19, 20, 23,** 24, **38**
    rubra, **23, 38**

Saccharum officinarum, 76
Salix spp., **20, 23,** 25, **39**
Secale cereale, 74
Stipa tenacissima, 77

Thuja plicata, **14,** 25, **36**
Tilia americana, **19, 20, 21,** 25, **39**
Triticum sativum, 74
Tsuga spp., **16**
    canadensis, 25, **35**
    heterophylla, 25, **35**

Ulmus spp., 25
    americana, **19, 21, 39**

Zea mays, 77

www.ingramcontent.com/pod-product-compliance
Lightning Source LLC
Chambersburg PA
CBHW051353070526
44584CB00025B/3741